Springer Theses

Recognizing Outstanding Ph.D. Research

Aims and Scope

The series "Springer Theses" brings together a selection of the very best Ph.D. theses from around the world and across the physical sciences. Nominated and endorsed by two recognized specialists, each published volume has been selected for its scientific excellence and the high impact of its contents for the pertinent field of research. For greater accessibility to non-specialists, the published versions include an extended introduction, as well as a foreword by the student's supervisor explaining the special relevance of the work for the field. As a whole, the series will provide a valuable resource both for newcomers to the research fields described, and for other scientists seeking detailed background information on special questions. Finally, it provides an accredited documentation of the valuable contributions made by today's younger generation of scientists.

Theses may be nominated for publication in this series by heads of department at internationally leading universities or institutes and should fulfill all of the following criteria

- They must be written in good English.
- The topic should fall within the confines of Chemistry, Physics, Earth Sciences, Engineering and related interdisciplinary fields such as Materials, Nanoscience, Chemical Engineering, Complex Systems and Biophysics.
- The work reported in the thesis must represent a significant scientific advance.
- If the thesis includes previously published material, permission to reproduce this must be gained from the respective copyright holder (a maximum 30% of the thesis should be a verbatim reproduction from the author's previous publications).
- They must have been examined and passed during the 12 months prior to nomination.
- Each thesis should include a foreword by the supervisor outlining the significance of its content.
- The theses should have a clearly defined structure including an introduction accessible to new PhD students and scientists not expert in the relevant field.

Indexed by zbMATH.

Su Yong Go

Photochemical and Electrochemical Activation Strategies of C(sp^3)-Based Building Blocks for Organic Synthesis

Doctoral Thesis accepted by
Seoul National University, Seoul, Korea (Republic of)

Springer

Author
Dr. Su Yong Go
Department of Chemistry
Seoul National University
Seoul, Korea (Republic of)

Supervisor
Prof. Hong Geun Lee
Seoul National University
Seoul, Korea (Republic of)

ISSN 2190-5053 ISSN 2190-5061 (electronic)
Springer Theses
ISBN 978-981-99-8993-5 ISBN 978-981-99-8994-2 (eBook)
https://doi.org/10.1007/978-981-99-8994-2

This Springer imprint is published by the registered company Springer Nature Singapore Pte Ltd.
The registered company address is: 152 Beach Road, #21-01/04 Gateway East, Singapore 189721, Singapore

Paper in this product is recyclable.

Supervisor's Foreword

The identification and utilization of a right functional handle is a central theme of research in controlled activation of organic molecules. This is especially true for redox-based organic synthesis processes that can form a variety of chemical bonds. Carbocation is a universally utilized high-energy intermediate for chemical synthesis that makes the formation of challenging chemical bonds possible. The contribution of Dr. Su Yong Go enabled a convenient access of the carbocation intermediate by the identification of a novel and operationally useful functional handle, the organic trifluoroborate.

With the close collaboration with the research group of Professor Taek Dong Chung, Dr. Go has scrutinized the redox behavior of Carbon-Boron bond positioned at an sp^3-hybridized carbon center. Eventually electrochemically-mediated strategy for the formation of the high-energy species was realized with a completely unprecedented level of generality. Dr. Go has demonstrated that the developed synthetic platform enables the formation of a variety of carbon-heteroatom bonds under highly complex settings.

This thesis describes the detailed path of Dr. Go's journey to arrive at the identification of novel electrochemical reactivity of organoboron compounds. The finding should not only enable the redesign of synthetic planning for the formation of challenging chemical bonds, but also lead to the discovery of other synthetic strategies that would contribute to the expansion of functional chemical space of biomedical research.

Seoul, Korea (Republic of)
October 2023

Prof. Hong Geun Lee

Preface

$C(sp^2)$–$C(sp^3)$ and $C(sp^3)$–heteroatom (C, N, O, F, P, S, Cl, and Se) bond-forming reactions were developed via photoredox or electrochemical methods. Part I introduces $C(sp^2)$–$C(sp^3)$ bond-forming reactions enabled by photoredox/nickel dual catalysis. Chapter 1 summarizes the major concepts and strategies in photoredox/nickel dual catalysis along with representative achievements with an emphasis on $C(sp^2)$–$C(sp^3)$ bond formation. Chapter 2 includes the hydroalkylation of alkynes via photoredox mediated Ni/Ir dual catalysis, which produces trisubstituted alkenes as versatile synthetic building blocks. High regioselectivity and *E/Z selectivity* were achieved by introducing silyl groups that can provide steric and electronic effects to these selectivities with extensive opportunities for post-functionalization.

Part II comprises the development of $C(sp^3)$–heteroatom bond-forming reactions via the electrochemical activation of $C(sp^3)$–B bonds. Unlike conventional activation strategies, the carbocationic activation of alkylboron compounds is in its infancy because of the constraints enforced by the reaction parameters. Chapter 3 presents the history of $C(sp^3)$–B bond activation and recent achievements of $C(sp^3)$–heteroatom bond formation in electroorganic synthesis. Chapter 4 describes the introduction of heteroatoms into alkyl carbocations generated from alkylboron reagents via electrochemical activation. This method enabled sterically hindered $C(sp^3)$–heteroatom bond formations under mild conditions and was applied for the functionalization of versatile bioactive or complex molecules. The reaction mechanism was elucidated based on electrochemical investigations and mechanistic experiments such as substituent effects, carbocation rearrangements, and kinetic studies.

Seoul, Korea (Republic of) — Dr. Su Yong Go
October 2023

Parts of this thesis have been adapted from the following published articles co-written by the author:

Chapter 2: "Highly Regioselective and *E/Z* Selective Hydroalkylation of Ynone, Ynoate, and Ynamide via Photoredox Mediated Ni/Ir Dual Catalysis" Go, S. Y.; Lee, G. S.; Hong, S. H. *Org. Lett.* **2018**, *20*, 4691.

Chapter 4: "A Unified Synthetic Strategy to Introduce Heteroatoms via Electrochemical Functionalization of Alkyl Organoboron Reagents" Go, S. Y.; Chung, H.; Shin, S. J.; An, S.; Youn, J. H.; Im, T. Y.; Kim, J. Y.; Chung, T. D.; Lee, H. *J. Am. Chem. Soc.* **2022**, *144*, 9149. (S.Y.G. and H.C. contributed equally to this work.)

Contributions

This work is the result of extensive collaborations between the author and other researchers. The specific contributions of the author are outlined below.

Chapter 2: This work was carried out by the author. Dr. Geun Seok Lee (Hong lab, SNU) designed the reaction for the first time.

Chapter 4: The author conceived the project and identified the reactivity for the first time. The author and Dr. Hyunho Chung (Lee lab, SNU) developed the synthetic protocol and conducted experiments to demonstrate the substrate scope and synthetic applications. Dr. Samuel Jaeho Shin (Chung lab, SNU) and Sohee An (Chung lab, SNU) carried out voltammetry measurements. Cyclic voltammogram and $n–t$ plots using TEAM were obtained by Dr. Samuel Jaeho Shin and Ji Yong Kim (Chung lab, SNU). The author carried out the in-depth investigations of the reaction mechanism such as divided cell experiments, constant potential electrolysis, and kinetic studies. Dr. Hyunho Chung carried out the investigation of substituent effects in substituted aromatic rings. Carbocation rearrangements were observed by the author and Dr. Hyunho Chung. Alkyl trifluoroborate salts were prepared by the author, Dr. Hyunho Chung, Ju Hyun Youn (Lee lab, SNU), and Tae Yeong Im (Lee lab, SNU). The X-ray structure in this chapter was solved by Dr. Ha-Jin Lee (the Western Seoul Center of Korea Basic Science Institute).

Acknowledgments

The life in graduate school has been one of the most important experiences in my life. I would like to express my gratitude to everyone who has enabled me to write this thesis.

First of all, I would like to thank my advisors, Prof. Soon Hyeok Hong and Prof. Hong Geun Lee for their support throughout the integrated master's and doctoral program, which allowed me to conduct my research. My advisors have provided invaluable guidance and numerous opportunities for my scholarly growth over the course of five and a half years. I am also thankful to Prof. Soon Hyeok Hong, Prof. David Yu-Kai Chen, and Prof. Chulbom Lee for giving me great opportunities to participate in the internship program during my undergraduate years, allowing me to gain experience in organic chemistry. These experiences played a crucial role in helping me choose organic chemistry as my research area and acquire a diverse range of knowledge in the field.

My sincere gratitude is also due to Dr. Seoksun Kim, Dr. Geun Seok Lee, and Dr. Hoyoon Park for mentoring me as an undergraduate intern when I had little knowledge of laboratory experiments. I would also like to extend my gratitude to Dr. Geun Seok Lee for providing invaluable guidance when I wrote my first research paper on photochemical reactions.

I deeply appreciate Dr. Hyun-Ho Chung, who worked with me to set up the electro organic reaction facility, explore the electrochemical reactivity of organic boron compounds, and tackle various problems together. I also would like to acknowledge Ju Hyun Yun, Tae Yeong Im, Dr. Jaeho Shin, Sohee An, and Ji Yong Kim for the fruitful collaborations. I believe that our collaborative efforts in addressing various challenges have contributed significantly to our interesting research outcomes.

I would like to express my gratitude to members of my Graduate Advising Committee, Prof. Byeong Moon Kim, Prof. Chulbom Lee, Prof. Taek Dong Chung, Prof. Hyunwoo Kim, and Prof. Hong Geun Lee. Especially, my sincere gratitude is due to Prof. Taek Dong Chung for his advice on electrochemical analysis experiments and for granting permission to use his laboratory facilities.

I sincerely appreciate my parents, grandmother, aunt, and uncle for their unwavering love, unfailing support, tremendous trust, and encouragement throughout all these years. Finally, I want to thank everyone who has become a valuable part of my life.

Contents

List of Figures

List of Schemes

List of Tables

Part I
C(sp^2)–C(sp^3) Bond Formation Reactions Enabled by Photoredox/Nickel Dual Catalysis

Chapter 1
The Key Concepts and Strategy in Photoredox/Nickel Dual Catalysis and Application to C–C Bond Formation Reactions

1.1 Introduction

Trisubstituted enones are versatile synthetic building blocks used for the synthesis of natural products and pharmaceutical agents. Various methods such as aldol reactions, olefination, cross-coupling, Heck reactions, and the organometallic-reagent additions have been developed to synthesize trisubstituted enones. However, these approaches are relatively inefficient in the aspects of atom and step economy. Recently, visible light photoredox catalysis has been recognized as a powerful synthetic tool. This method has been shown to photocatalytically generate chlorine atoms. The generation of alkyl radicals via hydrogen atom transfer (HAT) by halogen radicals and their addition to transition metals have enabled the selective intermolecular addition of nucleophilic radicals to alkynes.

This chapter describes the principles of visible light photoredox catalysis and its combination with transition metal catalysis for the development of $C(sp^2)$–$C(sp^3)$ bond-forming reactions. The representative examples are summarized and discussed for the availability of HAT by chlorine atoms under mild conditions.

1.2 Key Concepts in Visible Light Photoredox Catalysis

Prior to the development of visible light photoredox catalysis, ultraviolet light irradiation and absorption were directly used on organic molecules for their structural transformation. Organic molecules were activated by high-energy photons, showing high reactivity and novel selectivity compared to thermal activation [1]. However, the

S. Y. Go, *Photochemical and Electrochemical Activation Strategies of C(sp3)-Based Building Blocks for Organic Synthesis*, Springer Theses,
https://doi.org/10.1007/978-981-99-8994-2_1

molecules faced undesired decompositions or side reactions owing to high-energy absorption. To overcome these problems, visible light photoredox catalysis was developed using a photosensitizer, which delivers light energy indirectly to organic molecules as they cannot directly absorb visible light [2]. Consequently, much milder reaction conditions are available for the indirect activation of the target molecule.

The basic principles of visible light photoredox catalysis are explained as follows (Fig. 1.1) [2b, c, 3]. The role of a photosensitizer, for example, an iridium polypyridyl complex (*fac*-Ir(ppy)$_3$), is related to its emission properties. An electron in one of the metal-centered t_{2g} orbitals of Ir(ppy)$_3$ is excited to a ligand-centered π^* orbital through the absorption of a photon in the visible region. The singlet excited state undergoes rapid intersystem crossing (ISC) through metal-to-ligand charge transfer (MLCT), yielding the lowest-energy triplet MLCT state. The emission to the singlet ground state is forbidden without the influence of spin–orbit coupling (SOC). Owing to the forbidden pathway, the long lifetime of the triplet state of Ir(ppy)$_3$ remains and efficiently furnishes the function of single-electron transfer (SET) to other molecules. As shown in Fig. 1.1, the excited triplet state of Ir(ppy)$_3$ has both a high-energy electron in the π^* orbital and a low-energy hole in the t_{2g} orbital, simultaneously acting as oxidizing and reducing agents. Therefore, there are two photocatalytic cycles: (a) oxidative quenching cycle and (b) reductive quenching cycle (Fig. 1.2). After the reduction or oxidation of the excited triplet state of Ir(ppy)$_3$, the oxidized or reduced iridium polypyridyl complex undergoes an opposite redox event to regenerate the very first excited triplet state of Ir(ppy)$_3$, which is the key to developing visible light photoredox catalysis.

1.3 Principles of Photoredox/Nickel Dual Catalysis

Photoredox catalysis has been recognized as a valuable tool for the redox process of organic substrates, and new catalytic reactions have been developed to merge photoredox catalysis with transition metal catalysis [4]. Consequently, the improvements in new metallaphotoredox strategies present new opportunities to obtain novel bond disconnections and unprecedented reactivity. In other words, the application of photoinduced SET or energy transfer (ET) to the modification of the electronic structure of the metal catalyst expands opportunities for high-valence and excited-state catalysis without using stoichiometric redox agents or high-energy-light irradiation. While traditional transition metal catalysis operates a Ni(n)–Ni(n+2) catalytic cycle via oxidative addition, transmetalation, and reductive elimination (Fig. 1.3) [5], a photoredox/Ni dual catalysis furnishes a novel reaction mechanism (Fig. 1.4).

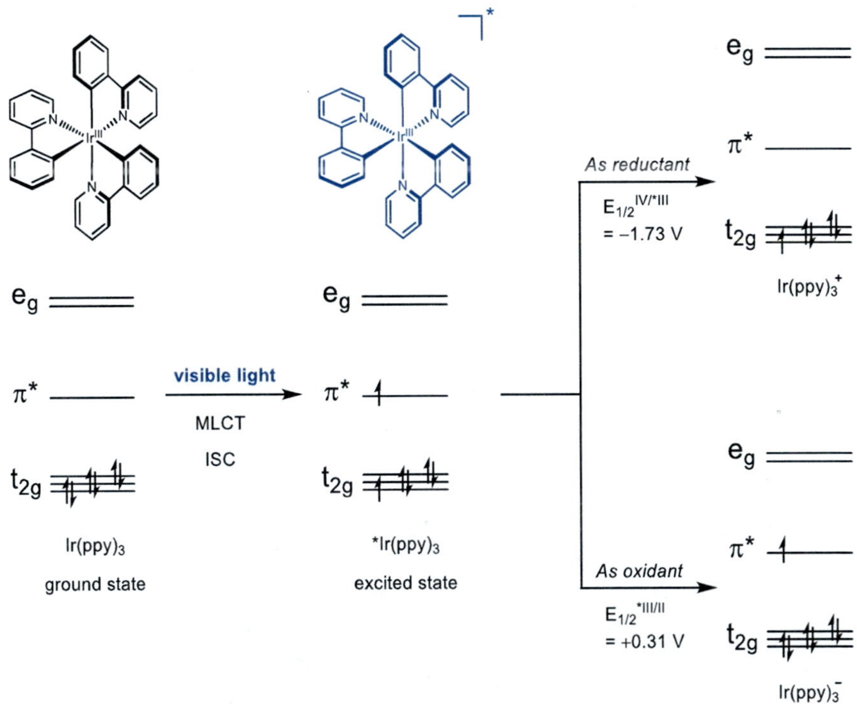

Fig. 1.1 Molecular orbital depiction of $Ir(ppy)_3$ photochemistry

Herein, photoredox catalysis is utilized to activate the nucleophilic compartment and complete the dual catalytic cycle by tuning the oxidation states of the Ni catalyst [6]. Generally, a Ni(0) species is inserted into the aryl electrophile, generating a Ni(II) species. Subsequently, the carbon-centered radical generated from an alkyl trifluoroborate upon SET with an excited Ir(III) complex adds to the Ni(II) species to produce a Ni(III)–alkyl intermediate. The electron-deficient Ni(III) is highly prone to reductive elimination, furnishing a Ni(I) intermediate. Finally, the Ni(I) species is reduced to the very first Ni(0) by an Ir(II) photocatalyst.

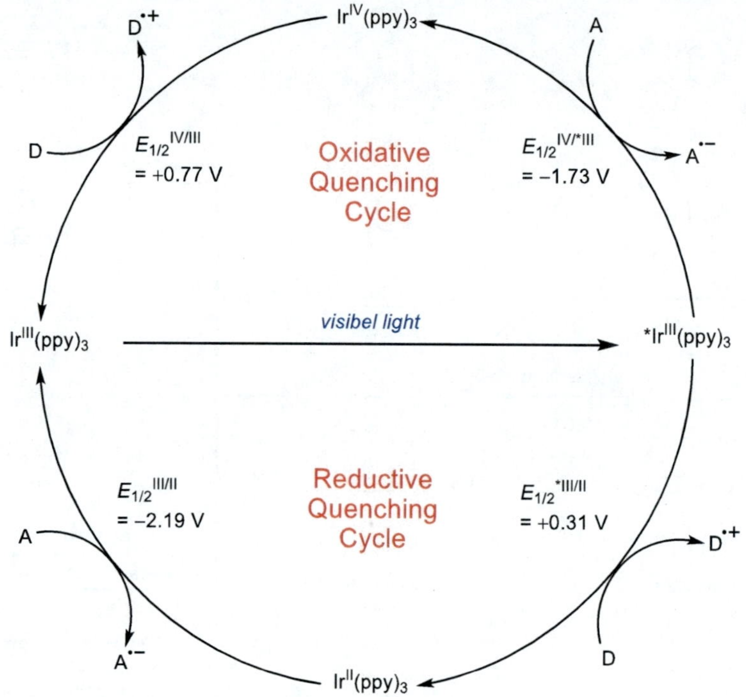

Fig. 1.2 Oxidative and reductive quenching cycle of Ir(ppy)$_3$

1.4 Photocatalytic C–H Activation and the Role of Chlorine Radicals

A direct oxidation approach is not available for C(sp^3)–H nucleophiles owing to their high oxidation potentials compared to C(sp^2)–H nucleophiles. In such a case, indirect radical C(sp^3)–H activation can be a feasible solution. Particularly, free-radical halogenation of alkanes is a well-known reaction in organic chemistry [7]. Particularly, chlorine radicals offer a wide opportunity to activate various C(sp^3)–H bonds by HAT. The HAT process is thermodynamically favorable, considering the bond-dissociation energies of the H–Cl bond (103 kcal/mol) and the C–H bonds (~100 kcal/mol) in unactivated alkanes [8].

Over the past decade, nickel catalysis has been used as a powerful platform for C(sp^3)–C bond formation [2b, 5, 6, 9]. In this context, HAT by halogen radicals has been integrated into Ni-catalyzed alkyl cross-coupling to achieve directing group-free C–C cross-coupling with even the strongest C(sp^3)–H bonds. Each method,

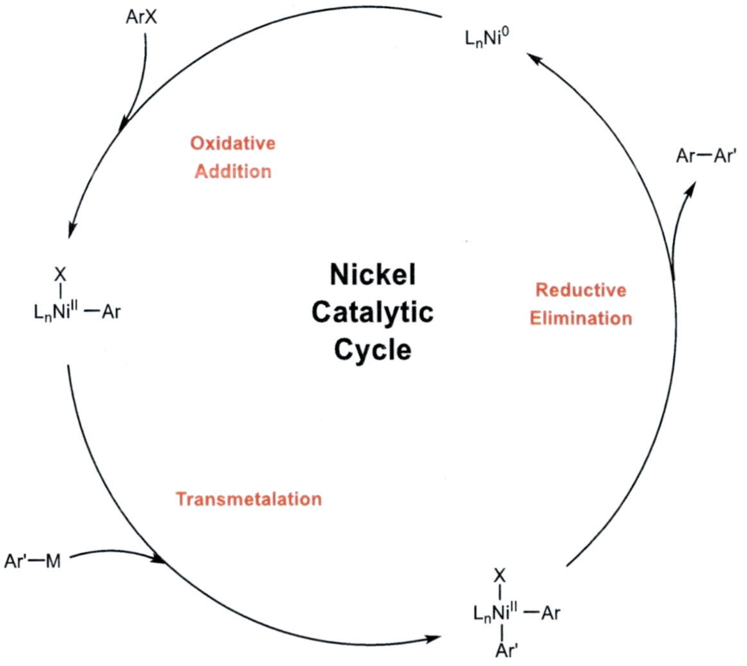

Fig. 1.3 A traditional nickel catalysis

developed by Doyle [10] and Molander [11] in 2016, is regarded as the first application of this integrated strategy to nickellaphotoredox C–C bond-forming reactions (Scheme 1.1).

A chlorine atom and a bromine atom generated from the homolytic cleavage of a Ni–X (X = Cl or Br) bond furnish α-heteroatom alkyl radicals enabled by HAT. The Doyle group described that a Ni(II) chloride intermediate is oxidized to a Ni(III) chloride complex by an excited Ir(III) photocatalyst (Scheme 1.1**A**). This Ni(III)–Cl bond is homolytically broken under light irradiation, as previously reported by the Nocera group [12]. Particularly, the reversible, visible-light-promoted elimination of chlorine atoms from Ni(III) trichloride complexes was reported by the Nocera group. Conversely, the Molander group revealed a different reaction mechanism in which a Ni(II) bromide intermediate is activated through ET with an excited Ir(III) photocatalyst (Scheme 1.1**B**). This process gives rise to an excited Ni(II) bromide species, eventually providing a bromine radical. The bromine radical would then undergo HAT against α-heteroatom alkyl compounds. After alkyl radicals are generated by Doyle's and Molander's methods, the subsequent radical addition and reductive elimination yield the desired products.

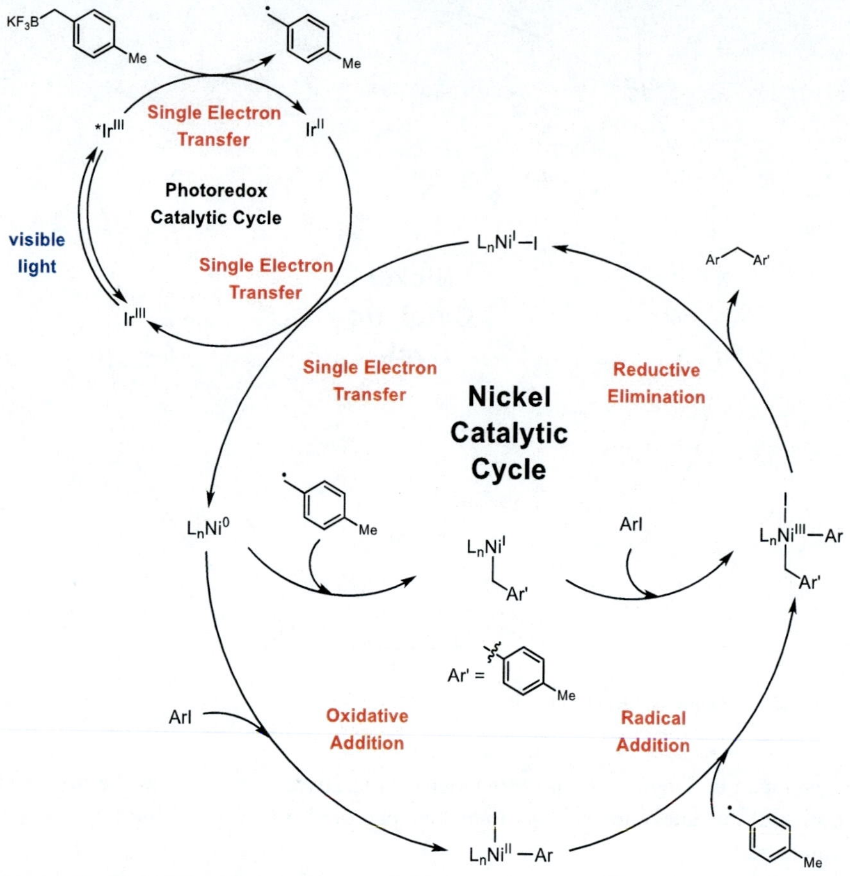

Fig. 1.4 A representative photoredox/nickel dual catalysis

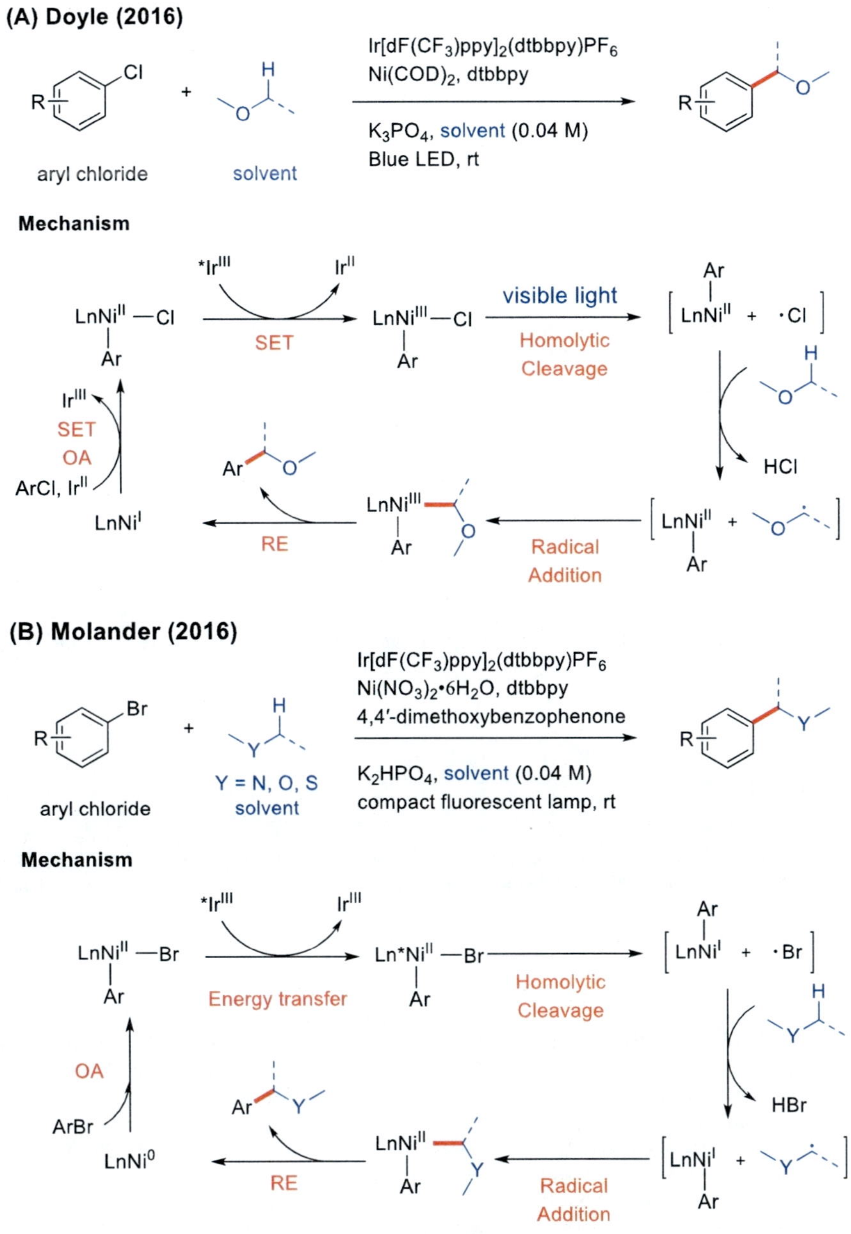

Scheme 1.1 Nickellaphotoredox catalysis via HAT by halogen radicals

1.5 Conclusion

The combination of transition metal and visible light photoredox catalysis has provided new opportunities to obtain novel bond disconnections and unprecedented reactivity under mild conditions. Owing to HAT using chlorine atoms, metallaphotoredox catalysis has been expanded to functionalize a variety of $C(sp^3)$–H substrates that cannot be directly oxidized.

For the growth of this field, further research directions should be focused on the expansion of the reaction scope by exploring the introduction of a Ni–alkyl intermediate to various reaction coupling partners. Enones are versatile synthetic building blocks used for the preparation of viable organic compounds. These substrates can be synthesized by the direct addition of alkyl radicals to carbon–carbon π bonds. However, the direct addition of radicals to alkynes is often problematic because of the sluggish rate of $C(sp^3)$–$C(sp^2)$ bond formation, low selectivity, and the generation of highly unstable vinyl radical intermediates that undergo undesirable open-shell pathways. As an alternative to naked alkyl radicals, the use of a Ni intermediate combined with an unstable alkyl radical might be a viable choice. In other words, the hydroalkylation of alkynes via migratory insertion pathway might provide a new strategy for the union of alkyl radicals and alkynes. Therefore, the ultimate goal of this activation strategy is the development of highly selective hydroalkylation of alkynes.

References

1. (a) Roth HD (1989) The beginnings of organic photochemistry. Angew Chem Int Ed 28:1193–1207. (b) Hoffmann N (2008) Photochemical reactions as key steps in organic synthesis. Chem Rev 108:1052–1103. (c) Bach T, Hehn JP (2011) Photochemical reactions as key steps in natural product synthesis. Angew Chem Int Ed 50:1000–1045. (d) Karkas MD, Porco JA Jr, Stephenson CR (2016) Photochemical approaches to complex chemotypes: applications in natural product synthesis. Chem Rev 116:9683–9747.
2. (a) Romero NA, Nicewicz DA (2016) Organic photoredox catalysis. Chem Rev 116:10075–10166. (b) Shaw MH, Twilton J, MacMillan DW (2016) Photoredox catalysis in organic chemistry. J Org Chem 81:6898–6926. (c) Prier CK, Rankic DA, MacMillan DW (2013) Visible light photoredox catalysis with transition metal complexes: applications in organic synthesis. Chem Rev 113:5322–5363.
3. (a) Hofbeck T, Yersin H (2010) The triplet state of fac-Ir(ppy)$_3$. Inorg Chem 49:9290–9299. (b) Koike T, Akita M (2014) Visible-light radical reaction designed by Ru- and Ir-based photoredox catalysis. Inorg Chem Front 1:562–576.
4. (a) Twilton J, Le C, Zhang P, Shaw MH, Evans RW, MacMillan DWC (2017) The merger of transition metal and photocatalysis. Nat Rev Chem 1:0052. (b) Zhu C, Yue H, Chu L, Rueping M (2020) Recent advances in photoredox and nickel dual-catalyzed cascade reactions: pushing the boundaries of complexity. Chem Sci 11:4051–4064. (c) Chan AY, Perry IB, Bissonnette NB, Buksh BF, Edwards GA, Frye LI, Garry OL, Lavagnino MN, Li BX, Liang Y, Mao E, Millet A, Oakley JV, Reed NL, Sakai HA, Seath CP, MacMillan DWC (2022) Metallaphotoredox: the merger of photoredox and transition metal catalysis 122:1485–1542.

5. Han F-S (2013) Transition-metal-catalyzed Suzuki-Miyaura cross-coupling reactions: a remarkable advance from palladium to nickel catalysts. Chem Soc Rev 42:5270–5298
6. Tellis JC, Primer DN, Molander GA (2014) Single-electron transmetalation in organoboron cross-coupling by photoredox/nickel dual catalysis. Science 345:433–436
7. (a) Fokin AA, Schreiner PR (2002) Selective alkane transformations via radicals and radical cations: insights into the activation step from experiment and theory. Chem Rev 102:1551–1594. (b) Carey FA, Sundberg RJ (2007) Advanced organic chemistry part A: structure and mechanisms. Springer, New York, p 965–1063
8. Blanksby SJ, Ellison GB (2003) Bond dissociation energies of organic molecules. Acc Chem Res 36:255–263
9. (a) Ananikov VP (2015) Nickel: the "spirited horse" of transition metal catalysis. ACS Catal 5:1964–1971. (b) Weix DJ (2015) Methods and mechanisms for cross-electrophile coupling of Csp^2 halides with alkyl electrophiles. Acc Chem Res 48:1767–1775. (c) Cornella J, Edwards JT, Qin T, Kawamura S, Wang J, Pan C, Gianatassio R, Schmidt M, Eastgate MD, Baran PS (2016) Practical Ni-catalyzed aryl–alkyl cross-coupling of secondary redox-active esters. J Am Chem Soc 138:2174−2177.
10. Shields BJ, Doyle AG (2016) Direct $C(sp^3)$–H cross coupling enabled by catalytic generation of chlorine radicals. J Am Chem Soc 138:12719–12722
11. Heitz DR, Tellis JC, Molander GA (2016) Photochemical nickel-catalyzed C-H arylation: synthetic scope and mechanistic investigations. J Am Chem Soc 138:12715
12. (a) Esswein AJ, Nocera DG (2007) Hydrogen production by molecular photocatalysis. Chem Rev 107:4022–4047. (b) Hwang SJ, Powers DC, Maher AG, Anderson BL, Hadt RG, Zheng SL, Chen S, Nocera DG (2015) Trap-free halogen photoelimination from mononuclear Ni(III) complexes. J Am Chem Soc 137:6472–6475. (c) Hwang SJ, Anderson BL, Powers DC, Maher AG, Hadt RG, Nocera DG (2015) Halogen photoelimination from monomeric nickel(III) complexes enabled by the secondary coordination sphere. Organometallics 34:4766–4774

Chapter 2
Highly Regioselective and *E/Z*-Selective Hydroalkylation of Alkyne via Photoredox-Mediated Ni/Ir Dual Catalysis

2.1 Introduction

Trisubstituted enones are versatile materials for the synthesis of pharmaceutical agents and natural products. A wide range of strategies such as aldol reactions [1], olefination [2], cross-coupling [3, 4], Heck reactions [5], C–H activation [6], and the addition of organometallic reagents [7] have been developed to synthesize trisubstituted enones. Still, these methods are less efficient, considering atom and step economy.

The direct, regioselective addition of C–H bonds to alkynes can serve as a step- and atom-economical reaction to give multi-substituted olefinic structures. In 2014, the Kang group reported the Co-catalyzed β-alkylation of terminal alkynes with tetrahydrofuran (THF) [8]. On the other hand, Hilt and co-workers reported a Zn-mediated α-selective addition of THF across aryl-propiolates, activated internal alkynes [9]. However, two methods suffered from low *E*/*Z* selectivity and limited substrate scope, showing that only terminal alkynes or propiolate derivatives could be applied. In 2015, a visible-light-promoted β-selective alkenylation of THF with propiolate derivatives was reported by the Wang group [10]. Still, this strategy faced the same problems, low *E*/*Z* selectivity and limited substrate scope, as only propiolate derivatives could be employed. When we were preparing the manuscript for publication, the Wu group presented the regioselective and *E*/*Z*-selective hydroalkylation of nonactivated alkynes with ethers via a visible-light-mediated Ni/Ir dual catalysis [11]. However, for an activated alkyne, ethyl 3-phenylpropiolate (one example), their synthetic platform failed to control the regioselectivity, affording α- and β-alkylated mixtures (1:1.5). Therefore, it is still challenging to generate highly α- and *E/Z*-selective trisubstituted enones by the direct addition of C—H bonds to activated alkynes (Scheme 2.1).

For the development of highly regioselective and *E*/*Z*-selective C–H addition to alkynes, two fundamental strategies were conceived. First, the introduction of a silyl group to alkynes, which can provide steric and electronic effects to control

S. Y. Go, *Photochemical and Electrochemical Activation Strategies of C(sp3)-Based Building Blocks for Organic Synthesis*, Springer Theses,
https://doi.org/10.1007/978-981-99-8994-2_2

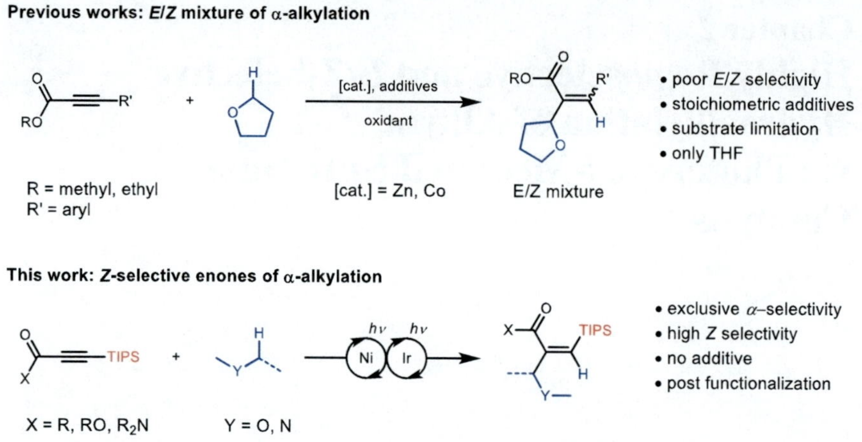

Scheme 2.1 Synthesis of trisubstituted enones through hydroalkylation of activated alkynes with C(sp^3)–H bonds

the reaction selectivity with opportunities for post-functionalization, was devised. Second, metallaphotoredox catalysis, which generates catalytically active alkyl radicals without any external radical sources, was adopted to exclude a stoichiometric amount of additives as employed in the recently reported C–H additions to alkynes [11, 12]. By realizing these strategies, a silyl-group-assisted C(sp^3)–H bond addition across alkynes was established via visible-light-mediated Ni/Ir dual catalysis to provide trisubstituted enones with exclusive *α*- and high *E*/*Z* selectivity. This synthetic protocol could be complementary to the previously reported *β*-alkylation strategies [10].

2.2 Results and Discussion

2.2.1 Optimization of Reaction Conditions

The initial screening test of reaction conditions was conducted with the reaction of ynones **1** with THF (Table 2.1). The evaluation of several silyl groups in compound **1** was performed to find the optimal, selectivity-inducing group. Silyl groups such as trimethylsilyl (TMS) and dimethylphenylsilyl (DMPS) groups afforded various unidentified products (entries 1 and 4). Slightly more stable silyl groups such as *tert*-butyldimethylsilyl (TBDMS) and triethylsilyl (TES) groups showed better yields and selectivities (entries 2 and 3). Introduction of the triisopropylsilyl (TIPS) group to ynone **1** exhibited the best yield (83%) with remarkable selectivities (entry 5). Notably, the *Z*-selective, *α*-addition product was preferred to the generally favored *E*-isomer, which originated from the reaction mechanism. The structure of compound

2 was identified by nuclear Overhauser effect (NOE) experiments (Figs. 2.2, 2.3, 2.4, 2.5, 2.6). Replacing TIPS group with aryl groups or less sterically hindered alkyl groups caused negative effects on selectivity or yield (Table 2.4). For instance, linear alkyl groups in ynone **1** resulted in a mixture of all four isomers (Table 2.4, entry 2). It could be explained that the bulkier TIPS group in compound **1** could make the reaction sterically sensitive, providing enough stability to give the desired products [13]. Substituting $NiCl_2$ with other Ni(II) or Ni(0) precatalysts exhibited significantly lower yields, showing that the role of chloride was essential for the desired transformation (entries 6 and 8). The combination of LiCl with $NiBr_2$ increased the product yield, confirming the critical role of the chloride ligand (entry 7). Conducting the reaction at high temperature (about 60 °C) resulted in significantly lower yields (entry 9). Finally, control experiments indicated the essential roles of the Ni catalyst, ligand, photocatalyst, and visible light (entry 10).

2.2.2 Substrate Scope

With the optimized reaction conditions in hand, the reactivity scope was investigated. The reactions proceeded efficiently with a variety of activated alkynes such as ynone, ynoate, and ynamide to achieve exclusive α-functionalization and high *E*/*Z* selectivity (Table 2.2), which overcame the limitations of the previous methods [8–11]. All compounds **1** afforded highly *Z*-selective, α-addition products and the *E*/*Z* isomer ratio of products (mostly >20:1, *Z*/*E*) was observed by ^{1}H NMR spectroscopy [14]. The reactions with various aliphatic TIPS-protected ynones (**2a–2e**) occurred smoothly under the optimized reaction conditions. On the other hand, the aromatic ynones exhibited lower yields. The reaction was facile with an electron-rich aromatic ynone (**2f**), while electron-poor aromatic ynones resulted in poor yields. The yield of benzyl ynone (**2g**) was increased by 30% compared to the phenyl analogue. To our delight, a ynoate (**2h**), a good precursor for various trisubstituted enones (Scheme 2.2), was a highly viable substrate for the transformation (98%). This strategy has also been applied to other bulkier ynoates and an ynamide (**2i–2l**).

Subsequently, different C–H reaction counterparts were investigated. Other cyclic and acyclic ethers, and even an amide could participate in hydroalkylation of alkynes. The role of benzene with a decreased amount of C–H coupling partners (20 equiv) was essential as a cosolvent for desired bond formation, owing to the solubility of the catalyst mixture (Table 2.3). In the cases of substrates **2o** and **2q**, the product yields were improved by increasing the reaction temperature (50 °C). 1,3-dioxolane underwent the desired transformation to show a slightly low yield, likely due to the decomposition of the product (**2p**). Interestingly, the activation of benzylic C(sp^3)–H bond could be accessed by utilizing an increased amount (80 mol %) of the Ni complex (**2r**). A ynoate **1h** also reacted with α-amino C(sp^3)–H bonds in dimethylacetamide (DMA) in a highly straightforward manner, leading to a high yield (**2s**).

Table 2.1 Optimization of THF alkenylation[a]

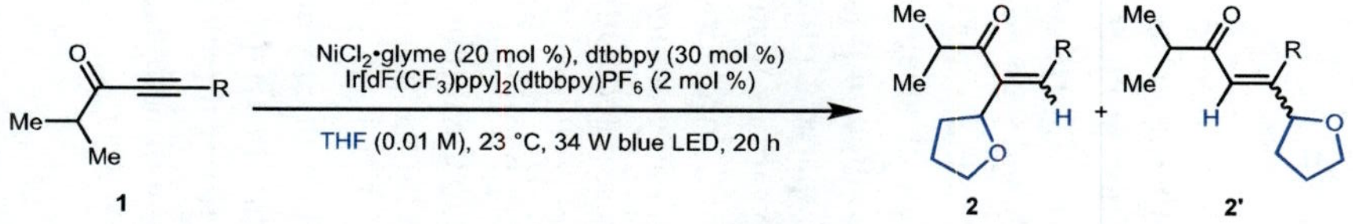

Entry	R	Deviation	Yield (%)[b] (**2**:**2'**), (*Z*:*E* of **2**)[c]
1	TMS	None	Complex mixture
2	TBDMS	None	52 (–), (1.6:1)
3	TES	None	48 (–), (3:1)
4	DMPS	None	Complex mixture
5	TIPS	None	83 (78[d]) (only **2**), (>20:1)
6	TIPS	$NiBr_2$·glyme instead of $NiCl_2$·glyme	18 (only **2**), (>20:1)
7	TIPS	$NiBr_2$·glyme + LiCl (1 equiv) instead of $NiCl_2$·glyme	41 (only **2**), (>20:1)
8	TIPS	$Ni(acac)_2$ or $Ni(cod)_2$instead of $NiCl_2$·glyme	0

(continued)

Table 2.1 (continued)

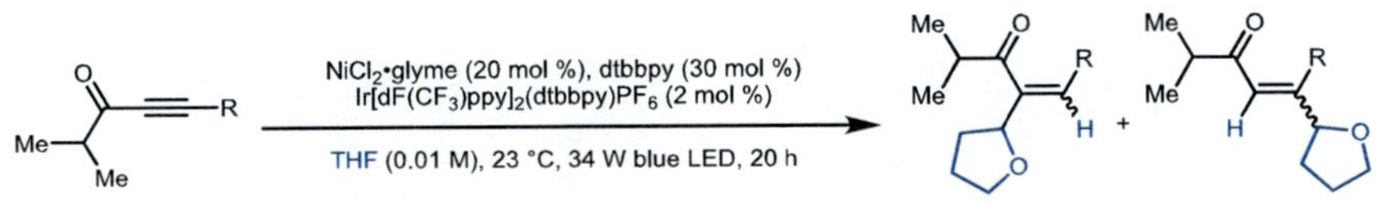

1 → 2 + 2'

Entry	R	Deviation	Yield (%)[b] (**2**:**2'**), (*Z*:*E* of **2**)[c]
9	TIPS	no cooling (60 °C) instead of 23 °C	18 (only **2**), (>20:1)
10	TIPS	without $NiCl_2$ or dtbbpy or [Ir] or light	0

[a]**1** (0.06 mmol), $NiCl_2$·glyme (20 mol %), dtbbpy (30 mol %), $Ir[dF(CF_3)ppy]_2(dtbbpy)PF_6$ (2 mol %), and THF (6.0 mL) irradiated with a 34 W blue LED
[b]NMR yields were calculated using 2,4,6-triiodophenol as the internal standard
[c]Selectivity ratios were determined by the crude ^1H NMR spectra
[d]Isolated yield

Table 2.2 Scope of ynone, ynoate, and ynamide[a]

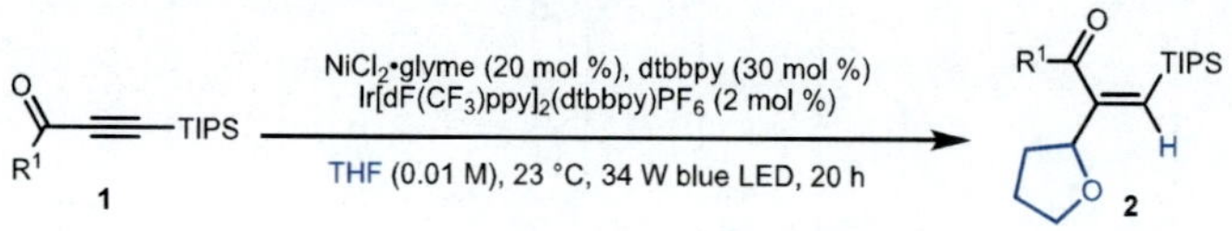

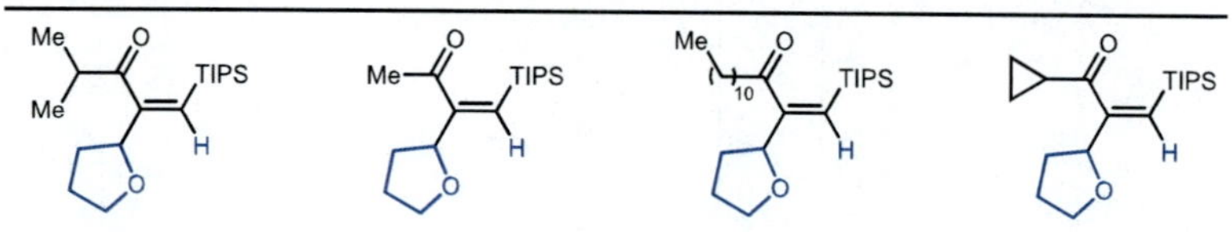

2a	**2b**	**2c**	**2d**
78%	61%	67%	67%

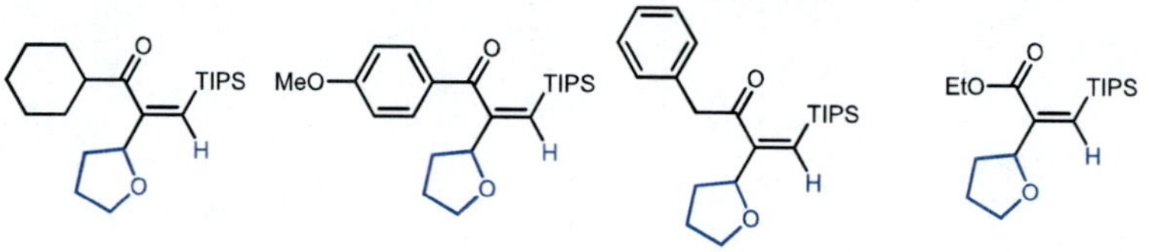

2e	**2f**	**2g**	**2h**
40%	55%	57%	98%

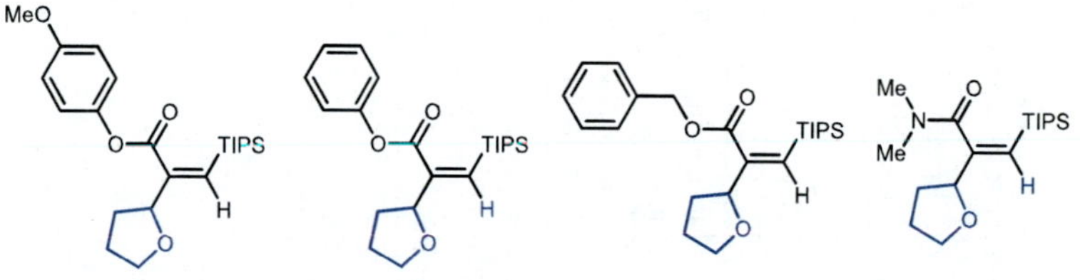

2i	**2j**	**2k**	**2l**
40%	54%[b]	70%	69%

[a]**1** (0.06 mmol), $NiCl_2$·glyme (20 mol %), dtbbpy (30 mol %), Ir[dF(CF_3)ppy]$_2$(dtbbpy)PF_6 (2 mol %), and THF (6.0 mL) irradiated with a 34 W blue LED. *Z*:*E* = >20:1, unless otherwise noted; Isolated yields of the *Z* isomer

[b] *Z*/*E* = 10:1

2.2.3 *Post-Functionalization and Mechanistic Studies*

Post-functionalization of enone **2h** was demonstrated, taking advantage of the TIPS group (Scheme 2.2). The silyl group was successfully converted into a halogen group under the halogenative desilylation conditions reported for nonactivated vinyl silanes [15]. The iodinated and brominated products (**3a, 3b**) were conveniently obtained in moderate yields, enabling additional transformations involving vinyl halides. For instance, the generated vinyl iodide **3a** could undergo arylation, maintaining *E*/*Z*-stereochemistry (**4a**) (Fig. 2.7). The synthetic utility of ester or Weinreb amide was also demonstrated by the preparation of allylic alcohol (**3c**) and the corresponding ynone (**2b**) [16].

Table 2.3 Scope of C(sp^3)–H partners in hydroalkylation[a]

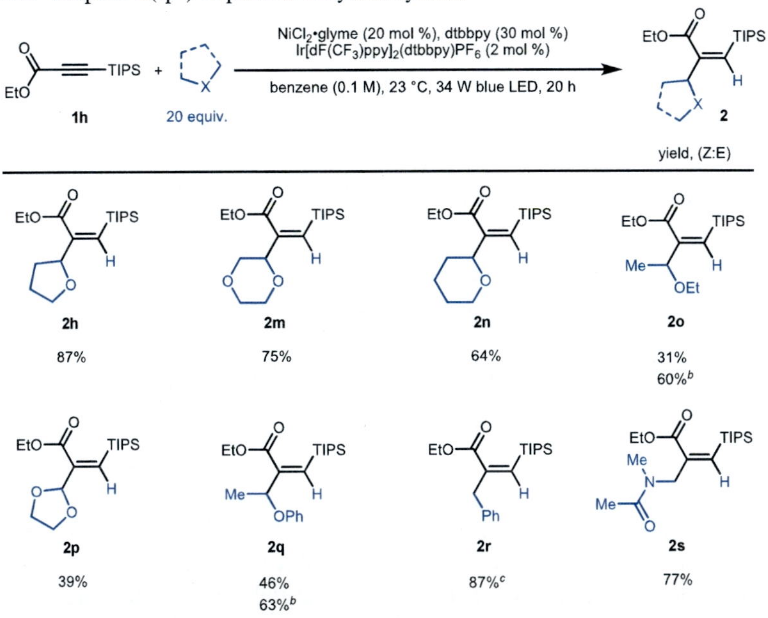

[a]Alkyne (0.10 mmol), $NiCl_2$·glyme (20 mol %), dtbbpy (30 mol %), Ir[dF(CF_3)ppy]$_2$(dtbbpy)PF_6 (2 mol %), and C–H donor (20 equiv), benzene (1.0 mL) irradiated with a 34 W blue LED. Z:E = >20:1; Isolated yields of the Z isomer

[b]50 °C

[c] $NiCl_2$·glyme (80 mol %), dtbbpy (90 mol %)

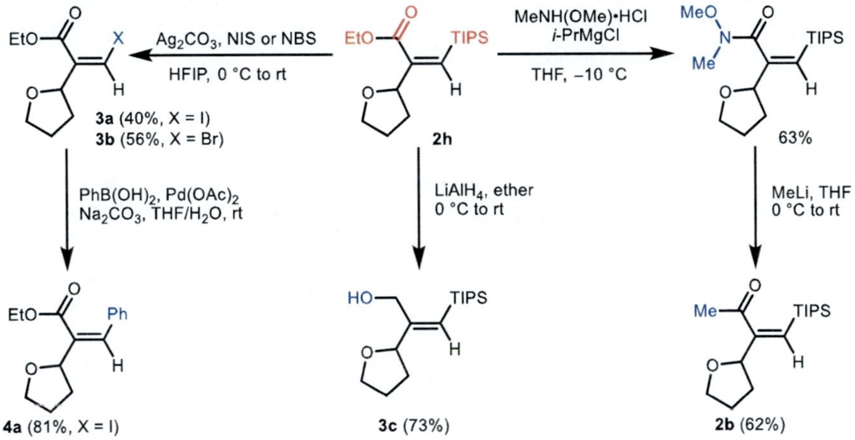

Scheme 2.2 Post-functionalization of TIPS enones

A proposed mechanism was presented based on control experiments (Scheme 2.3) and existing literature (Fig. 2.1) [11, 12]. As shown in Table 2.1, chloride turned out to be an essential component of the reaction system (entries 5–8). The Doyle group described that the photolysis of Ni(III)Cl_2 **B** could generate the chlorine radical, which could then abstract a hydrogen atom from THF to give an alkyl radical [12a,b]. The formation of the THF–TEMPO adduct in a radical quenching study also supported the generation of THF radical intermediates (Scheme 2.3a). The THF radical generated by a hydrogen atom transfer (HAT) process may rebound to Ni(II) species **C** and form the THF-bound Ni(III) complex **D**. The alkyl–Ni(III) complex **D** can be reduced to alkyl–Ni(II) species **E**. This alkyl–Ni(II) intermediate **E** undergoes 1,2-insertion with **1** to provide vinyl Ni complex **F**, as reported by the MacMillan group [12d]. The regioselectivity of the insertion process, as described by the Bergman group [17], and the steric effect of TIPS group result in the single isomer. The protodenickelation of the intermediate **F** would provide the desired product to regenerate the initial Ni(II) complex **A**, completing the catalytic cycle. Such a stepwise mechanism is consistent with deuterium labeling studies, showing that the vinyl hydrogen does not originate from THF (Figs. 2.9, 2.10, 2.11).

Based on experimental investigations, we suggested that this reaction proceeds through a Ni−alkyl pathway, not a Ni–hydride pathway. First, less sterically hindered ^{n}butyl ynoate was used to observe the trend in regioselectivity depending on the C–H

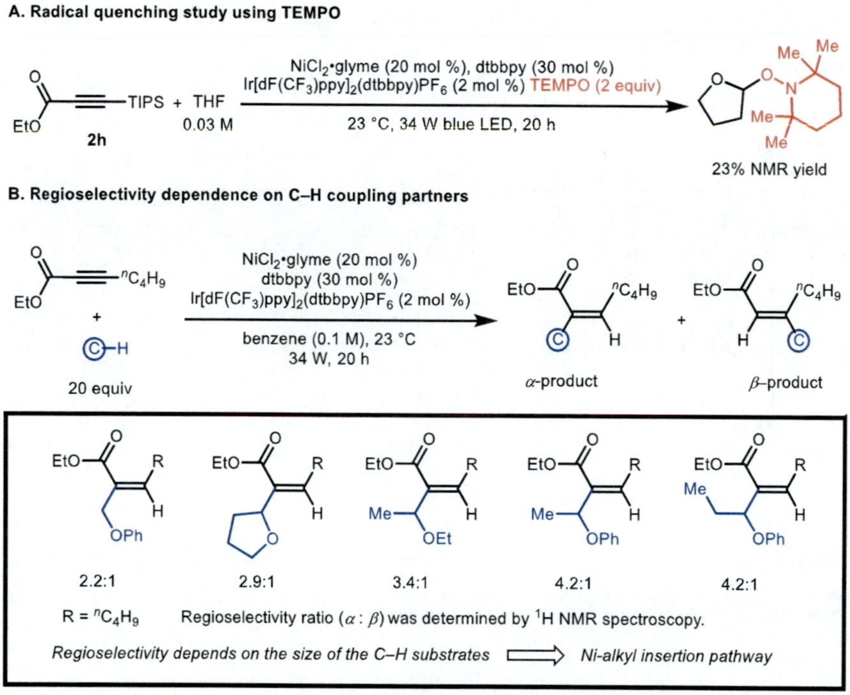

Scheme 2.3 Experimental investigations for determining the reaction mechanism

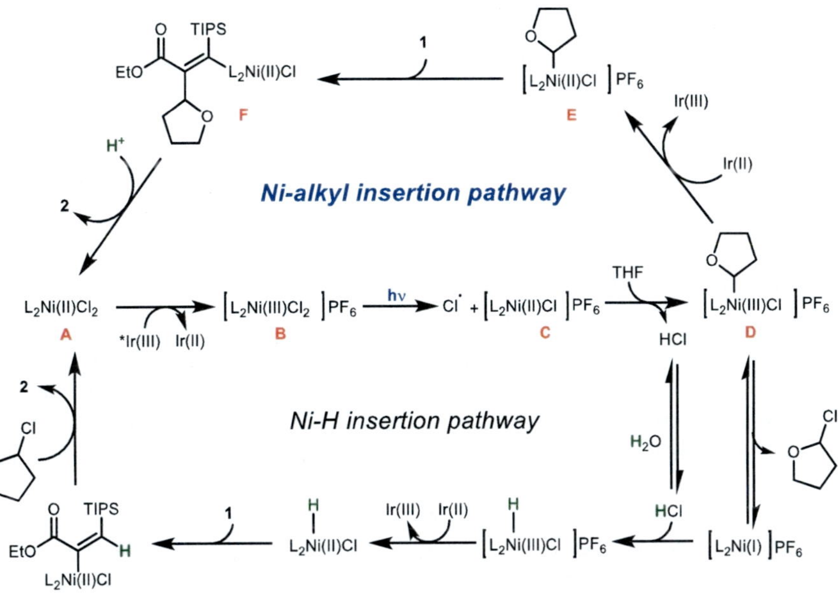

Fig. 2.1 Proposed mechanism

coupling partners, as silyl ynoates exhibited exceedingly high selectivity (Tables 2.3 and 2.6). The regioselectivity is increased with sterically bulkier C–H coupling partners consistent with a Ni–alkyl pathway (Scheme 2.3b). If the Ni–H pathway would be operative, the regioselectivity ratio should not be changed depending on the sterics of the C–H reaction counterparts, owing to the 1,2-insertion of the same Ni–H species to the alkynes regardless of the counterparts [12d]. Second, the reaction was carried out in good yields (50%–60%) even with the addition of bases, although a base can inhibit the generation of a nickel hydride intermediate by quenching HCl (Fig. 2.1, Table 2.7).

2.3 Conclusion

In summary, a TIPS-group-assisted regioselective and *E*/*Z*-selective hydroalkylation of alkynes was developed for the synthesis of trisubstituted enones via photoredox-mediated Ni/Ir dual catalysis. The reaction afforded the α-selective *Z* isomer with highly controlled selectivity endowed by the steric properties of the TIPS group. As a result, this synthetic protocol is complementary to the β-alkylation process via nucleophilic Giese reactions and other hydroalkylation reactions of alkynes. Post-functionalization of the products could give diverse trisubstituted alkenes. A proposed mechanism involves the catalytic generation of alkyl radicals and 1,2-insertion of a nickel–alkyl intermediate into alkynes, which sequentially undergoes protodemetalation to complete the hydroalkylation reaction.

2.4 Experimental Section

2.4.1 General Information

Unless otherwise noted, all reactions were performed under inert conditions. Nuclear magnetic resonance (NMR) spectra were recorded in $CDCl_3$ on a Bruker DPX-300 (300 MHz) spectrometer or Varian 400 and 500 NMR (400 and 500 MHz) spectrometers, with the residual solvent signal was used as a reference. Liquid chromatography-mass spectrometry (LC–MS) spectra were obtained on an Agilent 6120 Quadrupole LC/MS. High-resolution mass spectrometry (HRMS) was performed at the Organic Chemistry Research Center in Sogang University using the electrospray ionization (ESI) method. Chemical shifts are reported in ppm and coupling constants are given in Hz. Reactions were monitored by thin-layer chromatography (TLC) on EMD Silica Gel 60 F254 plates, and visualized either using UV light (254 nm) or by staining with potassium permanganate and heating. Tetrahydrofuran (THF) and benzene were dried using a PureSolv solvent purification system. Deuterated compounds were purchased from Cambridge Isotope Laboratories, Inc. and Sigma-Aldrich Corporation. For all visible-light photocatalytic reactions, 34 W blue LED lamps purchased from Kessil (Kessil H150 Grow Light-Blue) were used for all the visible-light photocatalytic reactions.

2.4.2 Substrate Preparation

All alkynes were prepared according to a previously reported literature procedure with slight modifications. The photocatalyst Ir[dF(CF_3)ppy]$_2$(dtbbpy)PF_6 was synthesized as described previously [18].

General Procedure A

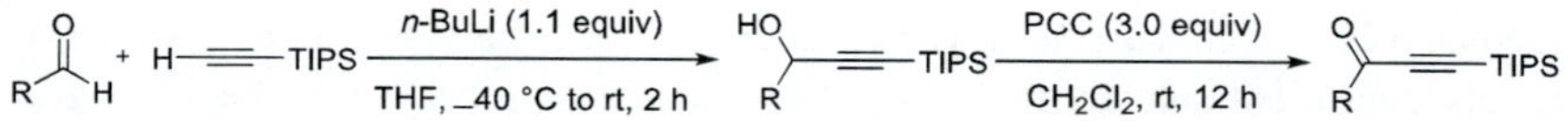

Alkynes (**1a–1f**) were synthesized via a previously reported procedure with slight modifications [19]. To a stirred solution of triisopropyl(prop-1-yn-1-yl)silane (1.00 g, 5.5 mmol) in anhydrous THF (20 mL), ^{n}BuLi (1.6 M in ^{n}hexane, 3.4 mL, 5.5 mmol)

was added at –40 °C under Ar. The reaction mixture was then stirred at the same temperature for 1 h. After 1 h, subsequently, the corresponding aldehyde (5.0 mmol) dissolved in anhydrous THF (5 mL) was added dropwise to the reaction mixture at –40 °C. The resulting mixture was stirred for 1 h at –40 °C. The resulting mixture was warmed to room temperature for 2 h before quenching with NH_4Cl (sat. aq, 20 mL). The layers were separated and the aqueous layer was extracted with EtOAc (3 × 10 mL). The combined organic extracts were washed with brine, dried (anhydrous Na_2SO_4), and concentrated *in vacuo* to afford the alcohol that was pure enough to be used directly in the next step. To a stirred solution of the corresponding alcohol in CH_2Cl_2 (20 mL), pyridinium chlorochromate (3.23 g, 15.0 mmol) was added at room temperature. The resulting mixture was stirred for 12 h before it was filtered through a pad of Celite. The filtrate was concentrated *in vacuo* and purified by flash column chromatography (silica gel, hexanes/CH_2Cl_2, gradient elution) to afford the desired product.

General Procedure B

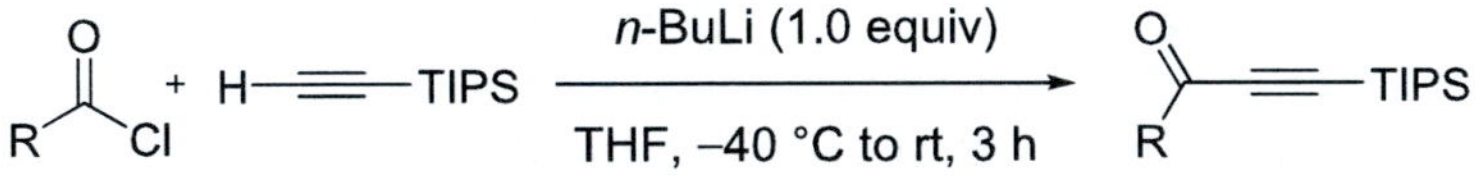

Alkynes (**1h–1l**) were synthesized via a previously reported procedure with slight modifications [20]. To a stirred solution of triisopropyl(prop-1-yn-1-yl)silane (0.91 g, 5.0 mmol) in anhydrous THF (20 mL), ^{n}BuLi (1.6 M in ^{n}hexane, 3.1 mL, 5.0 mmol) was added at –40 °C under Ar. The reaction mixture was then stirred at the same temperature for 1 h. After 1 h, subsequently, the corresponding acyl halide (5.0 mmol) in anhydrous THF (5 mL) was added dropwise to the reaction mixture at –40 °C. The resulting mixture was stirred for 1 h at the same temperature and then warmed to room temperature. It was stirred for another 2 h before it was quenched with NH_4Cl (sat. aq, 20 mL). The layers were separated and the aqueous layer was extracted with EtOAc (3 × 10 mL). The combined organic extracts were washed with brine, dried (anhydrous Na_2SO_4), and concentrated *in vacuo*. The resulting residue was purified by flash column chromatography (silica gel, hexanes/CH_2Cl_2, gradient elution) to afford the desired product.

General Procedure C

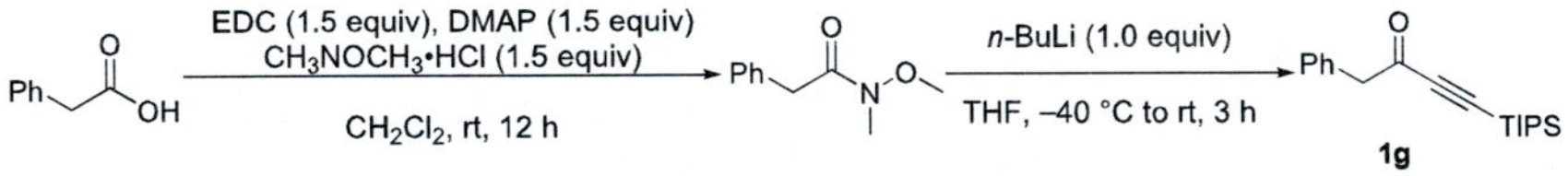

1g

Alkyne **1g** was synthesized via a previously reported procedure with slight modifications [21]. To a stirred solution of the corresponding carboxylic acid (0.68 g, 5.0 mmol) in anhydrous CH_2Cl_2 (25 mL), 1-ethyl-3-(3-dimethylaminopropyl)carbodiimide (1.44 g, 7.5 mmol), 4-dimethylaminopyridine (0.92 g, 7.5 mmol), and *N,O*-dimethylhydroxylamine hydrochloride (0.73 g, 7.5 mmol) were added at room temperature. The reaction mixture was then stirred at the same temperature for 12 h before it was quenched with NH_4Cl (sat. aq, 25 mL). The layers were separated and the aqueous layer was extracted with CH_2Cl_2 (3 × 10 mL). The combined organic extracts were washed with brine, dried (anhydrous Na_2SO_4), and concentrated *in vacuo* to afford the corresponding Weinreb amide. These amides were pure enough to be used directly in the next step. To a stirred solution of triisopropyl(prop-1-yn-1-yl)silane (0.91 g, 5.0 mmol) in anhydrous THF (20 mL), ^{n}BuLi (1.6 M in ^{n}hexane, 3.1 mL, 5.0 mmol) was added at –40 °C under Ar. Next, the reaction mixture was stirred at the same temperature for 1 h. Subsequently, the corresponding Weinreb amide in anhydrous THF (5 mL) was added dropwise to the reaction mixture at –40 °C. The resulting mixture was stirred for 1 h and then warmed to room temperature. It was stirred for another 2 h before it was quenched with NH_4Cl (sat. aq, 25 mL). The layers were separated and the aqueous layer was extracted with EtOAc (3 × 10 mL). The combined organic extracts were washed with brine, dried (anhydrous Na_2SO_4), and concentrated *in vacuo*. The resulting residue was purified by flash column chromatography (silica gel, hexanes/CH_2Cl_2, gradient elution) to afford **1g** (0.80 g, 2.7 mmol, 53% yield).

Previously reported alkyne substrates: 4-Methyl-1-(triisopropylsilyl) pent-1-yn-3-one (**1a**) [22], 4-(triisopropylsilyl)but-3-yn-2-one (**1b**) [23], 1-(triisopropylsilyl)tetradec-1-yn-3-one (**1c**) [24], 1-cyclopropyl-3-(triisopropylsilyl) prop-2-yn-1-one (**1d**) [25], 1-cyclohexyl-3-(triisopropylsilyl)prop-2-yn-1-one (**1e**) [22], 1-(4-methoxyphenyl)-3-(triisopropylsilyl)prop-2-yn-1-one (**1f**) [23], 1-phenyl-4-(triisopropylsilyl)but-3-yn-2-one (**1g**) [22], ethyl 3-(triisopropylsilyl) propiolate (**1h**) [26], *N,N*-dimethyl-3-(triisopropylsilyl)propiolamide (**1l**) [23] were prepared according to the general procedure. ^{1}H and ^{13}C NMR spectra data were matched with those reported in the literature.

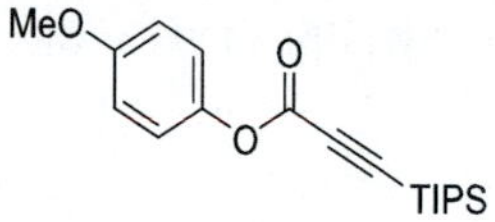

4-Methoxyphenyl 3-(Triisopropylsilyl)propiolate (1i)

Colorless oil, 0.59 g (1.77 mmol, 35% yield); ^{1}H NMR (300 MHz, $CDCl_3$): δ = 7.10–7.06 (m, 2H), 6.93–6.88 (m, 2H), 3.81 (s, 3H), 1.18–1.08 (m, 21H); ^{13}C NMR

(75 MHz, $CDCl_3$): δ = 157.6, 151.6, 143.6, 122.2, 114.5, 96.0, 94.3, 55.5, 18.4, 10.9; HRMS-ESI (*m/z*) $[M+Na]^+$ calcd for $C_{19}H_{28}NaO_3Si$, 355.1700; found: 355.1701.

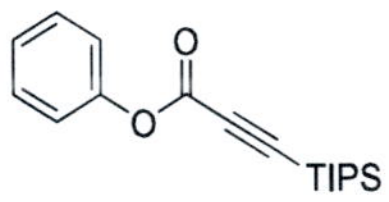

Phenyl 3-(Triisopropylsilyl)propiolate (1j)
Yellow oil, 0.65 g (2.15 mmol, 43% yield); ^1H NMR (300 MHz, $CDCl_3$): δ = 7.40 (t, J = 7.8 Hz, 2H), 7.26 (t, J = 7.4 Hz, 1H), 7.16 (d, J = 8.0 Hz, 2H), 1.13 (d, J = 5.2 Hz, 21H); ^{13}C NMR (75 MHz, $CDCl_3$): δ = 151.1, 150.2, 129.5, 126.3, 121.5, 96.0, 94.5, 18.4, 11.0; HRMS-ESI (*m/z*) $[M+Na]^+$ calcd for $C_{18}H_{26}NaO_2Si$, 325.1594; found: 325.1595.

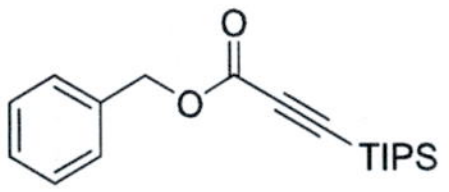

Benzyl 3-(Triisopropylsilyl)propiolate (1k)
Colorless oil, 1.1 g (3.48 mmol, 70% yield); ^1H NMR (300 MHz, $CDCl_3$): δ = 7.43–7.37 (m, 5H), 5.22 (s, 2H), 1.15–1.08 (m, 21H); ^{13}C NMR (75 MHz, $CDCl_3$): δ = 152.9, 135.0, 128.6, 128.5, 128.5, 96.5, 91.8, 67.4, 18.4, 10.9; HRMS-ESI (*m/z*) $[M+Na]^+$ calcd for $C_{19}H_{28}NaO_2Si$, 339.1751; found: 339.1752.

2.4.3 *Reaction Optimization*

See Tables 2.4 and 2.5.

2.4.4 *Identification of α/β- and E/Z Configuration of Products*

The *E/Z* configuration of compounds **2d, 2h–j,** and **2l** is identified according to the nuclear Overhauser effect (NOE) in ^1H NMR (Figs. 2.2, 2.3, 2.4, 2.5, 2.6, 2.7).

Table 2.4 Investigation of reaction selectivity

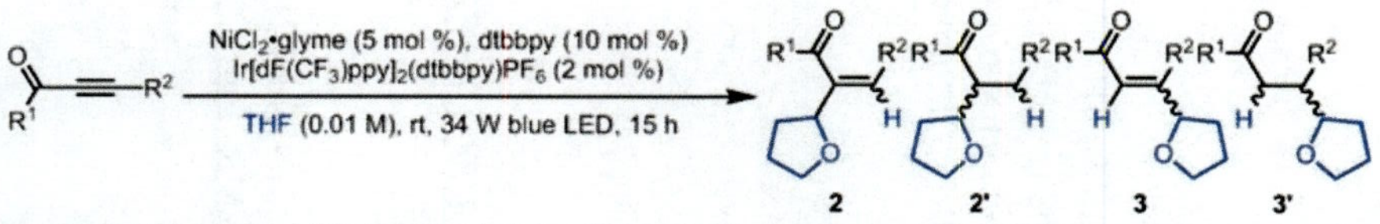

Entry	R^1	R^2	Yield (%)	2: 2': 3: 3'
1	*p*-MeOPh	Ph	22	0: 0: 2.3: 1
2	*i*-propyl	*n*-C_4H_9	52	1.1: 0: 1: 0
3	*i*-propyl	cyclohexyl	–	Complex mixture
4	*i*-propyl	*p*-MeOPh	–	Complex mixture
5	*i*-propyl	TIPS	34	1: 0: 0: 0

Table 2.5 Optimization of the THF alkenylation

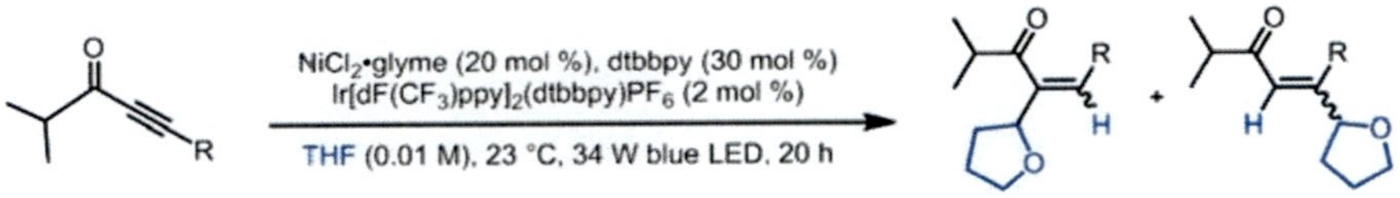

1 2 2'

Entry	R	Deviation	Yield (%)	2:2′	Z:E of 2
1	TMS	None	–	Complex mixture	
2	TES	None	52	–	1.6:1
3	TBDMS	None	48	–	3:1
4	TIPS	None	83	Only 2	>20:1
5	DMPS	None	–	Complex mixture	
6	TIPS	Without NiClj or dtbbpy or Ir[dF(CF_3)ppy]$_2$(dtbbpy)PF_E or light	0	Only 2	>20:1
7	TIPS	$NiBr_2$-glyme instead of $NiCl_2$-glyme	18	Only 2	>20:1
8	TIPS	$NiBr_2$-glyme + LiCI (1 equiv) instead of $NiCl_2$-glyme	41	Only 2	>20:1
9	TIPS	Ni(acac)$_2$ or Ni(cod)$_2$ instead of $NiCl_2$-glyme	0	Only 2	>20:1
10	TIPS	Ru(bpy)$_3^{2t}$ or Eosin Y or Mes-Acr^{4}" instead of Ir[dF(CF_3)ppy]$_2$(dtbbpy)PF_6	0	Only 2	>20:1
11	TIPS	Adding $HC0_2H$ (1 equiv)	6	Only 2	>20:1
12	TIPS	Adding K_2C0_3(1 equiv)	52	Only 2	>20:1
13	TIPS	Adding DBU (1 equiv)	0	Only 2	>20:1
14	TIPS	Adding pyridine (1 equiv)	34	Only 2	>20:1

(continued)

Table 2.5 (continued)

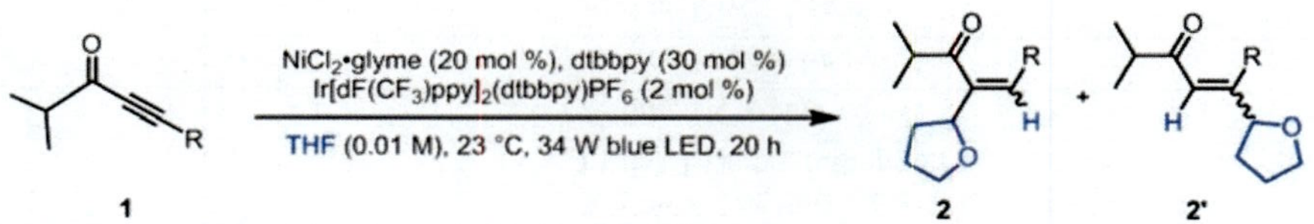

Entry	R	Deviation	Yield (%)	2:2′	Z:E of 2
15	TIPS	SIPr-HCI + K_2C0_3 (1 equiv) instead of dtbbpy	4	Only 2	>20:1
16	TIPS	BiOx instead of dtbbpy	7	Only 2	>20:1
17	TIPS	Bathophenanthroline instead of dtbbpy	32	Only 2	>20:1
18	TIPS	No cooling (60 °C) instead of room temperature	18	Only 2	>20:1

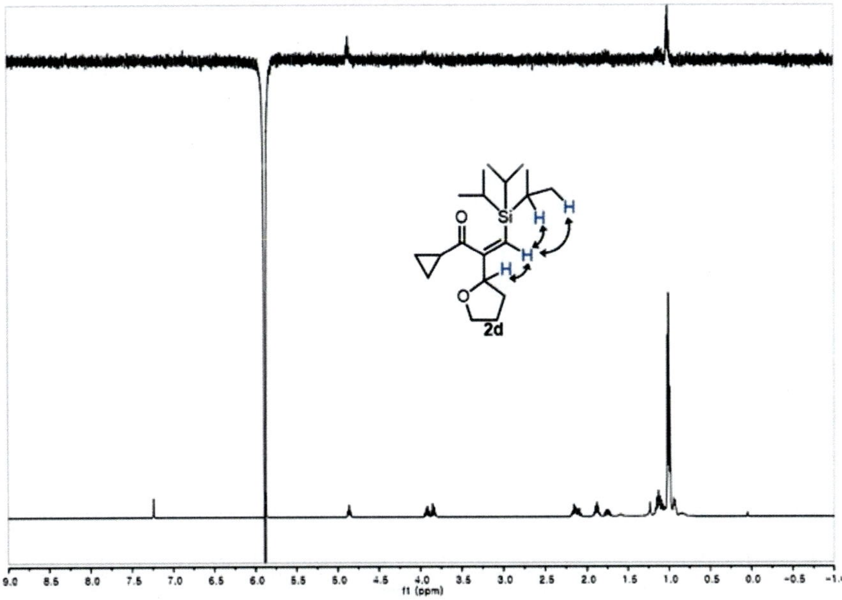

Fig. 2.2 ^{1}H NMR spectra of **2d** and NOE experiment

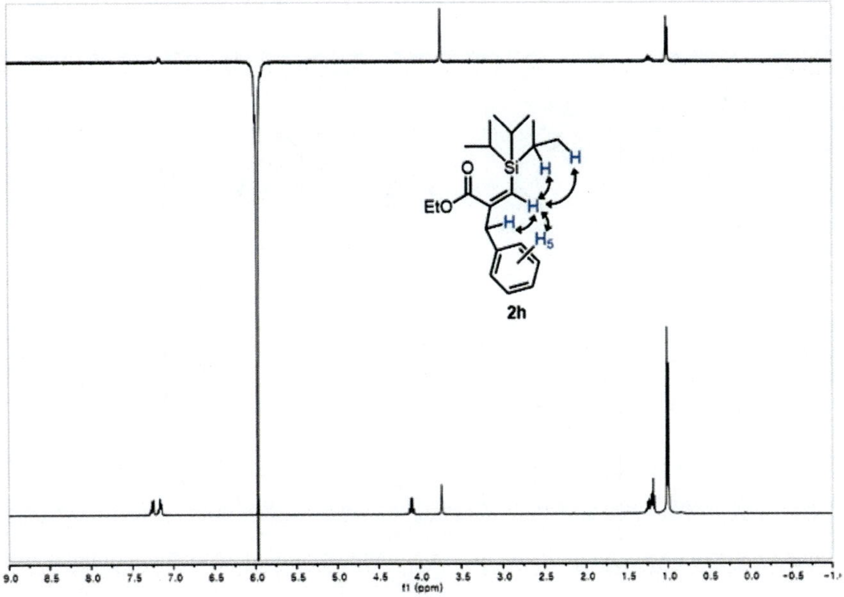

Fig. 2.3 ^{1}H NMR spectra of **2h** and NOE experiment

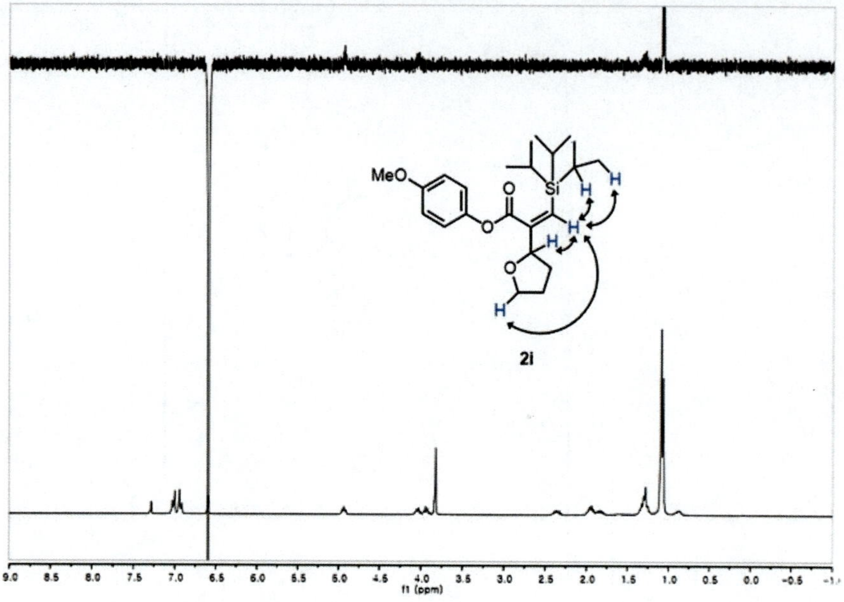

Fig. 2.4 ^{1}H NMR spectra of **2i** and NOE experiment

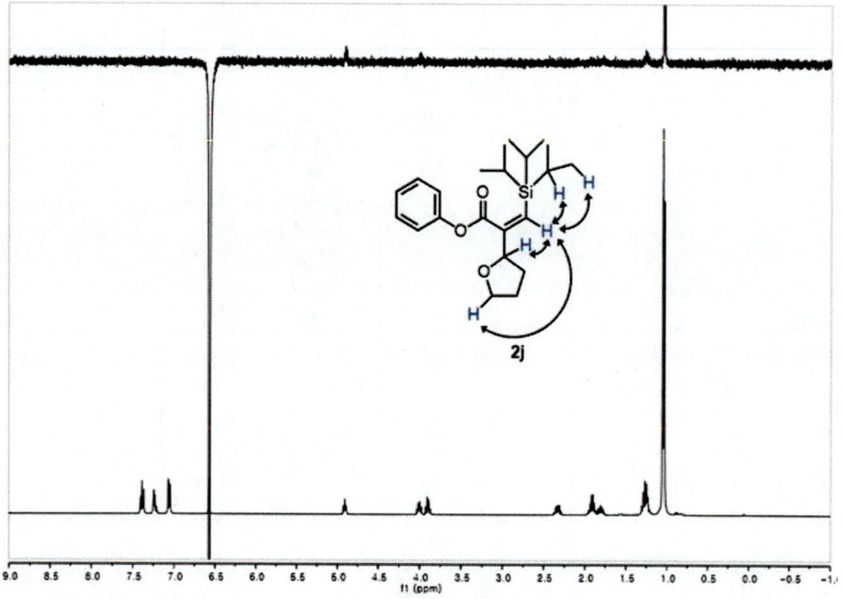

Fig. 2.5 ^{1}H NMR spectra of **2j** and NOE experiment

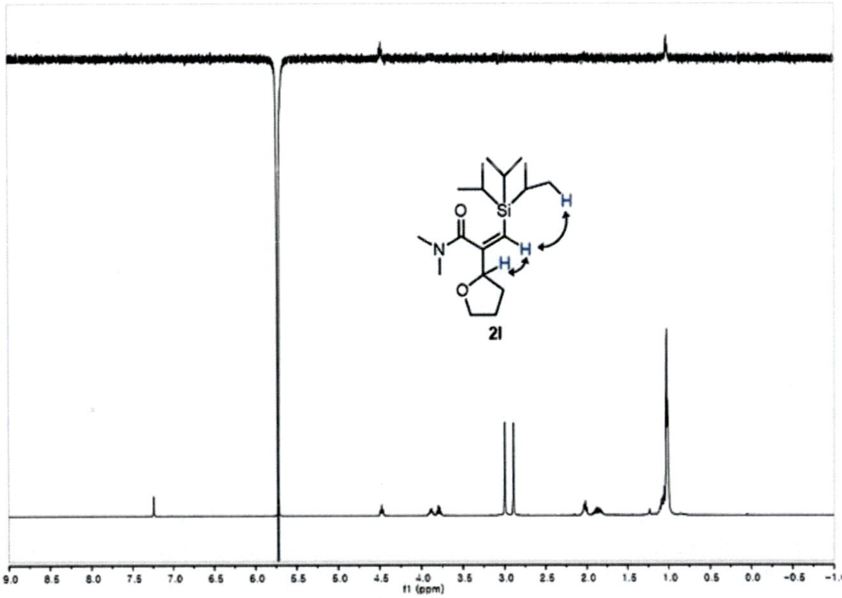

Fig. 2.6 ^{1}H NMR spectra of **2l** and NOE experiment

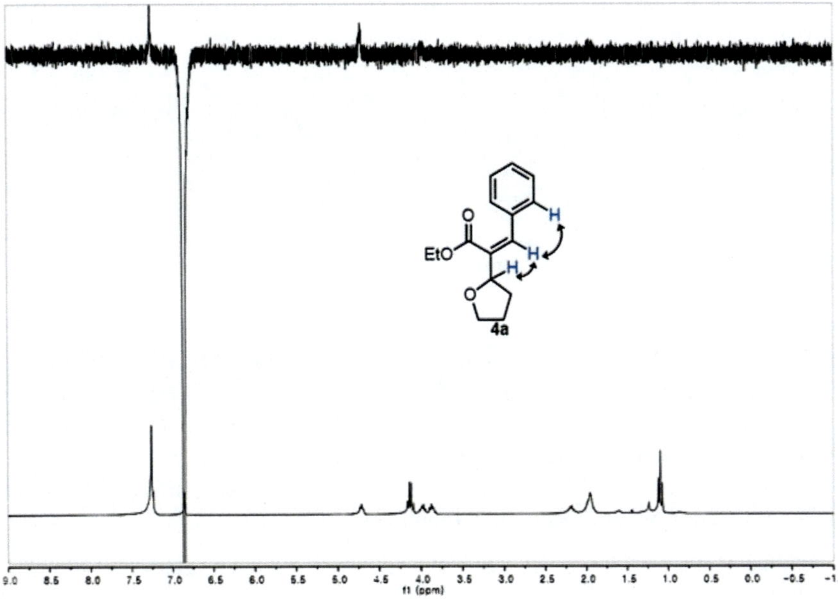

Fig. 2.7 ^{1}H NMR spectra of **4a** and NOE experiment

2.4.5 General Procedure for the Hydroalkylation of Activated Alkynes

General Procedure A

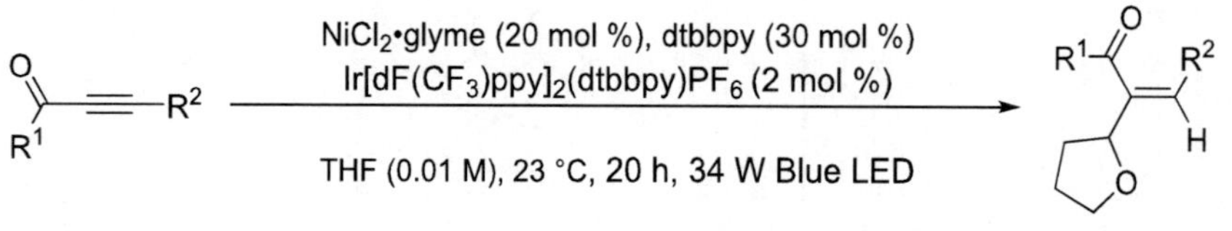

To a vial equipped with a PTFE-coated stirrer bar, $NiCl_2$·glyme (2.64 mg, 0.012 mmol), dtbbpy (4.83 mg, 0.018 mmol), and THF (6.0 mL) were added. The resulting solution was stirred for 5 min to give a green suspension before $Ir[dF(CF_3)ppy]_2(dtbbpy)PF_6$ (1.35 mg, 0.0012 mmol) and the corresponding alkyne (0.06 mmol) were added to the vial. The resulting mixture was stirred for 20 h under 34 W blue LED irradiation at 23 °C under Ar. Subsequently, the reaction mixture was concentrated *in vacuo* and purified by flash column chromatography (silica gel, hexanes/CH_2Cl_2, gradient elution or hexanes/EtOAc, gradient elution) to afford the desired product.

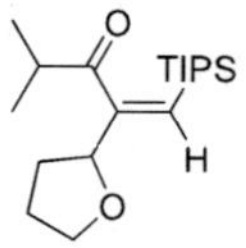

(Z)-4-Methyl-2-(tetrahydrofuran-2-yl)-1-(triisopropylsilyl)pent-1-en-3-one (2a)
Colorless oil, 15.2 mg (0.047 mmol, 78% yield); ^1H NMR (300 MHz, $CDCl_3$): δ = 6.09 (d, J = 1.4 Hz, 1H), 4.82 (ddd, J = 7.0, 5.7, 1.3 Hz, 1H), 3.99–3.92 (m, 1H), 3.89–3.82 (m, 1H), 2.92 (hept, J = 6.9 Hz, 1H), 2.22–2.09 (m, 1H), 1.95–1.84 (m, 2H), 1.74–1.63 (m, 1H), 1.26–1.16 (m, 3H), 1.12 (dd, J = 6.8, 6.1 Hz, 6H), 1.03 (dd, J = 7.2, 4.7 Hz, 18H); ^{13}C NMR (75 MHz, $CDCl_3$): δ = 207.8, 157.6, 131.1, 79.7, 68.4, 38.1, 32.2, 25.1, 19.1, 19.1, 18.6, 18.0, 12.2; HRMS-ESI (*m/z*) $[M+Na]^+$ calcd for $C_{19}H_{36}NaO_2Si$, 347.2377; found: 347.2378.

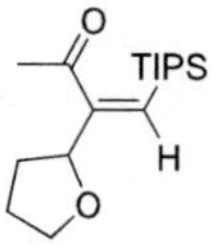

(Z)-3-(Tetrahydrofuran-2-yl)-4-(triisopropylsilyl)but-3-en-2-one (2b)
Colorless oil, 10.9 mg (0.037 mmol, 61% yield); ^{1}H NMR (300 MHz, $CDCl_3$): δ = 5.93 (d, J = 1.3 Hz, 1H), 4.77 (t, J = 6.5 Hz, 1H), 3.94 (dd, J = 13.9, 7.6 Hz, 1H), 3.85 (dd, J = 14.7, 7.6 Hz, 1H), 2.30 (s, 3H), 2.22–2.09 (m, 1H), 1.99–1.86 (m, 2H), 1.80–1.70 (m, 1H), 1.23–1.13 (m, 3H), 1.02 (dd, J = 7.1, 2.7 Hz, 18H); ^{13}C NMR (75 MHz, $CDCl_3$): δ = 202.5, 158.5, 129.6, 80.0, 68.3, 31.6, 29.5, 25.2, 19.0, 19.0, 12.0; HRMS-ESI (m/z) $[M+Na]^+$ calcd for $C_{17}H_{32}NaO_2Si$, 319.2064; found: 319.2066.

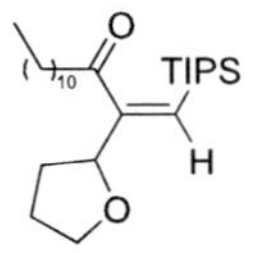

(Z)-2-(Tetrahydrofuran-2-yl)-1-(triisopropylsilyl)tetradec-1-en-3-one (2c)
Colorless oil, 17.5 mg (0.040 mmol, 67% yield); ^{1}H NMR (300 MHz, $CDCl_3$): δ = 5.90 (s, 1H), 4.78 (t, J = 6.5 Hz, 1H), 3.93 (dd, J = 14.3, 7.1 Hz, 1H), 3.84 (dd, J = 14.7, 7.5 Hz, 1H), 2.72–2.46 (m, 2H), 2.19–2.08 (m, 1H), 1.98–1.84 (m, 2H), 1.79–1.68 (m, 1H), 1.65–1.54 (m, 2H), 1.34–1.21 (m, 16H), 1.21–1.12 (m, 3H), 1.02 (dd, J = 7.0, 3.9 Hz, 18H), 0.88 (t, J = 6.6 Hz, 3H); ^{13}C NMR (75 MHz, $CDCl_3$): δ = 205.0, 158.9, 128.5, 80.0, 68.3, 41.8, 31.9, 31.7, 29.6, 29.5, 29.5, 29.4, 29.2, 25.2, 23.3, 22.7, 19.1, 19.0, 14.1, 12.0; HRMS-ESI (m/z) $[M+Na]^+$ calcd for $C_{27}H_{52}NaO_2Si$, 459.3629; found: 459.3630.

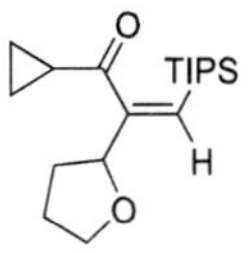

(Z)-1-Cyclopropyl-2-(tetrahydrofuran-2-yl)-3-(triisopropylsilyl)prop-2-en-1-one (2d)
Colorless oil, 12.9 mg (0.039 mmol, 67% yield); ^{1}H NMR (400 MHz, $CDCl_3$): δ = 5.91 (s, 1H), 4.89 (t, J = 6.5 Hz, 1H), 3.95 (dd, J = 14.1, 7.2 Hz, 1H), 3.87 (dd, J = 14.8, 7.3 Hz, 1H), 2.22–2.09 (m, 2H), 1.95–1.86 (m, 2H), 1.81–1.73 (m, J = 32.0 Hz, 1H), 1.20–1.08 (m, 5H), 1.06–1.01 (m, 18H), 1.00–0.94 (m, 2H); ^{13}C NMR (75 MHz, $CDCl_3$): δ = 206.2, 159.6, 126.8, 80.2, 68.4, 31.7, 25.3, 20.3, 19.0, 19.0, 12.9, 12.2, 11.9; HRMS-ESI (m/z) $[M+Na]^+$ calcd for $C_{19}H_{34}NaO_2Si$, 345.2220; found: 345.2222.

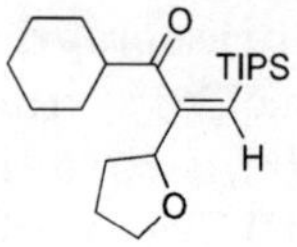

(Z)-1-Cyclohexyl-2-(tetrahydrofuran-2-yl)-3-(triisopropylsilyl)prop-2-en-1-one (2e)

Colorless oil, 8.7 mg (0.024 mmol, 40% yield); ^{1}H NMR (400 MHz, $CDCl_3$): δ = 6.07 (d, J = 1.3 Hz, 1H), 4.83–4.77 (m, 1H), 4.00–3.92 (m, 1H), 3.91–3.81 (m, J = 22.4 Hz, 1H), 2.70–2.56 (m, 1H), 2.24–2.08 (m, 1H), 1.96–1.78 (m, 6H), 1.75–1.64 (m, 2H), 1.33–1.14 (m, 8H), 1.03 (dd, J = 7.2, 4.5 Hz, 18H); ^{13}C NMR (75 MHz, $CDCl_3$): δ = 207.2, 157.8, 130.8, 79.8, 68.4, 48.4, 32.3, 28.9, 28.2, 26.0, 25.9, 25.8, 25.1, 19.2, 19.1, 12.3; HRMS-ESI (*m/z*) $[M+Na]^+$ calcd for $C_{22}H_{40}NaO_2Si$, 387.2690; found: 387.2693.

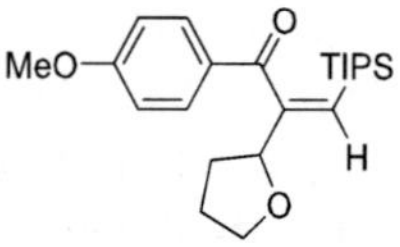

(Z)-1-(4-Methoxyphenyl)-2-(tetrahydrofuran-2-yl)-3-(triisopropylsilyl)prop-2-en-1-one (2f)

Colorless oil, 12.8 mg (0.033 mmol, 55% yield); ^{1}H NMR (300 MHz, $CDCl_3$): δ = 7.97–7.90 (m, 2H), 6.95–6.89 (m, 2H), 6.00 (d, J = 1.4 Hz, 1H), 4.74–4.68 (m, 1H), 3.93–3.83 (m, 4H), 3.82–3.75 (m, 1H), 2.05–1.77 (m, 4H), 1.14–1.04 (m, 3H), 1.00–0.92 (m, 18H); ^{13}C NMR (125 MHz, $CDCl_3$): δ = 198.2, 163.7, 159.1, 132.1, 130.2, 124.0, 113.6, 81.7, 68.6, 55.4, 31.7, 25.2, 18.9, 18.8, 11.7; HRMS-ESI (*m/z*) $[M+Na]^+$ calcd for $C_{23}H_{36}NaO_3Si$, 411.2326; found: 411.2328.

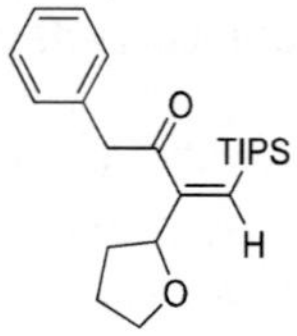

(Z)-1-Phenyl-3-(tetrahydrofuran-2-yl)-4-(triisopropylsilyl)but-3-en-2-one (2g)
Colorless oil, 12.8 mg (0.034 mmol, 57% yield); ^{1}H NMR (300 MHz, $CDCl_3$): δ = 7.36–7.16 (m, 5H), 5.98 (d, *J* = 1.4 Hz, 1H), 4.87–4.81 (m, 1H), 4.05–3.82 (m, 4H), 2.09 (ddd, *J* = 14.2, 11.7, 7.1 Hz, 1H), 1.96–1.84 (m, 2H), 1.80–1.67 (m, 1H), 1.24–1.11 (m, 3H), 1.03 (dd, *J* = 7.1, 3.4 Hz, 18H); ^{13}C NMR (75 MHz, $CDCl_3$): δ = 202.1, 158.5, 133.9, 129.9, 129.8, 128.5, 126.8, 80.1, 68.4, 48.3, 31.7, 25.2, 19.1, 19.0, 12.0; HRMS-ESI (*m/z*) $[M+Na]^+$ calcd for $C_{23}H_{36}NaO_2Si$, 395.2377; found: 395.2379.

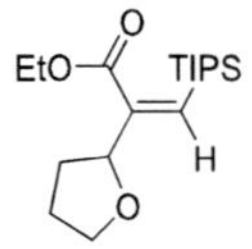

Ethyl (Z)-2-(Tetrahydrofuran-2-yl)-3-(triisopropylsilyl)acrylate (2h)
Yellow oil, 19.2 mg (0.059 mmol, 98% yield); ^{1}H NMR (300 MHz, $CDCl_3$): δ = 6.32 (d, *J* = 1.4 Hz, 1H), 4.78 (ddd, *J* = 7.2, 5.6, 1.4 Hz, 1H), 4.22 (q, *J* = 7.1 Hz, 2H), 4.01–3.94 (m, 1H), 3.92–3.82 (m, 1H), 2.28–2.17 (m, 1H), 1.95–1.81 (m, 2H), 1.76–1.64 (m, 1H), 1.34–1.21 (m, 6H), 1.04 (dd, *J* = 7.3, 1.8 Hz, 18H); ^{13}C NMR (75 MHz, $CDCl_3$): δ = 167.3, 150.4, 133.0, 79.4, 68.6, 60.4, 32.5, 25.2, 19.1, 19.0, 14.2, 12.2; HRMS-ESI (*m/z*) $[M+Na]^+$ calcd for $C_{18}H_{34}NaO_3Si$, 349.2169; found: 349.2168.

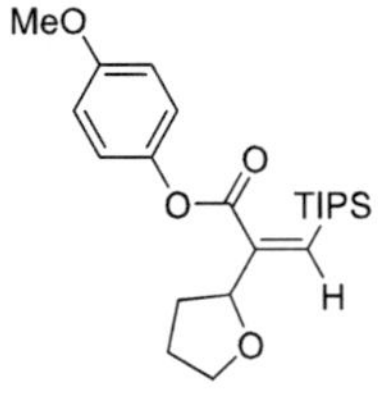

4-Methoxyphenyl (Z)-2-(Tetrahydrofuran-2-yl)-3-(triisopropylsilyl)acrylate (2i)
Colorless oil, 9.7 mg (0.024 mmol, 40% yield); ^{1}H NMR (300 MHz, $CDCl_3$): δ = 7.04–6.97 (m, 2H), 6.96–6.89 (m, 2H), 6.58 (s, 1H), 4.92 (dd, *J* = 7.1, 5.8 Hz, 1H), 4.03 (dd, *J* = 13.5, 7.3 Hz, 1H), 3.92 (dd, *J* = 14.5, 7.3 Hz, 1H), 3.81 (s, 3H), 2.41–2.28 (m, 1H), 2.00–1.88 (m, 2H), 1.88–1.76 (m, 1H), 1.34–1.23 (m, 3H), 1.07

(d, J = 7.2 Hz, 18H); ^{13}C NMR (75 MHz, $CDCl_3$): δ = 165.9, 157.3, 149.5, 143.8, 136.3, 122.3, 114.6, 79.4, 68.7, 55.6, 32.7, 25.3, 19.11, 19.06, 12.1; HRMS-ESI (*m/z*) $[M+Na]^+$ calcd for $C_{23}H_{36}NaO_4Si$, 427.2275; found: 427.2277.

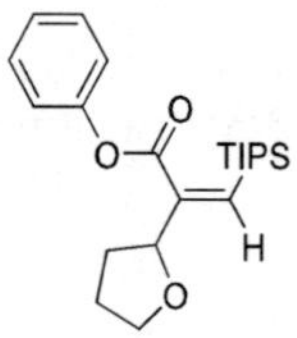

Phenyl (Z)-2-(Tetrahydrofuran-2-yl)-3-(triisopropylsilyl)acrylate (2j)
Colorless oil, 12.1 mg (0.032 mmol, 54% yield); ^{1}H NMR (300 MHz, $CDCl_3$): δ = 7.41 (t, J = 7.9 Hz, 2H), 7.26 (t, J = 7.4 Hz, 1H), 7.09 (d, J = 7.7 Hz, 2H), 6.60 (d, J = 1.3 Hz, 1H), 4.94 (ddd, J = 6.8, 5.5, 1.1 Hz, 1H), 4.03 (dd, J = 13.4, 7.5 Hz, 1H), 3.93 (dd, J = 15.0, 7.0 Hz, 1H), 2.41–2.30 (m, 1H), 2.00–1.90 (m, 2H), 1.89–1.78 (m, 1H), 1.38–1.23 (m, 3H), 1.07 (d, J = 7.2 Hz, 18H); ^{13}C NMR (75 MHz, $CDCl_3$): δ = 165.5, 150.4, 149.5, 136.6, 129.6, 125.9, 121.6, 79.4, 68.7, 32.7, 25.3, 19.1, 19.1, 12.1; HRMS-ESI (*m/z*) $[M+Na]^+$ calcd for $C_{22}H_{34}NaO_3Si$, 397.2169; found: 397.2169.

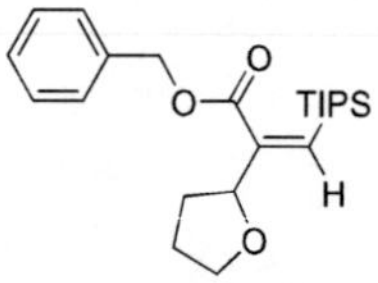

Benzyl (Z)-2-(Tetrahydrofuran-2-yl)-3-(triisopropylsilyl)acrylate (2k)
Colorless oil, 16.3 mg (0.042 mmol, 70% yield); ^{1}H NMR (300 MHz, $CDCl_3$): δ = 7.38–7.35 (m, 5H), 6.36 (d, J = 1.4 Hz, 1H), 5.20 (s, 2H), 4.80 (ddd, J = 7.1, 5.6, 1.3 Hz, 1H), 3.99–3.92 (m, 1H), 3.90–3.80 (m, 1H), 2.26–2.14 (m, 1H), 1.92–1.79 (m, 2H), 1.75–1.62 (m, 1H), 1.25–1.14 (m, 3H), 1.01 (dd, J = 7.3, 1.7 Hz, 18H); ^{13}C NMR (75 MHz, $CDCl_3$): δ = 167.1, 150.1, 135.7, 133.9, 128.5, 128.5, 128.3, 79.3, 68.6, 66.3, 32.6, 25.1, 19.0, 19.0, 12.1; HRMS-ESI (*m/z*) $[M+Na]^+$ calcd for $C_{23}H_{36}NaO_3Si$, 411.2326; found: 411.2328.

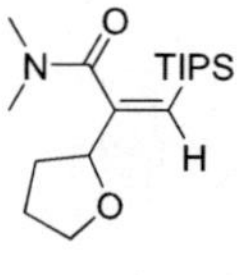

(Z)-*N*,*N*-Dimethyl-2-(tetrahydrofuran-2-yl)-3-(triisopropylsilyl)acrylamide (2l)

Colorless oil, 13.4 mg (0.041 mmol, 69% yield); ^{1}H NMR (300 MHz, $CDCl_3$): δ = 5.75 (d, *J* = 0.9 Hz, 1H), 4.50 (td, *J* = 6.6, 1.1 Hz, 1H), 3.91 (dd, *J* = 13.7, 7.4 Hz, 1H), 3.82 (dd, *J* = 14.3, 7.5 Hz, 1H), 3.03 (s, 3H), 2.92 (s, 3H), 2.10–2.00 (m, 2H), 1.97–1.79 (m, 2H), 1.13–1.00 (m, 21H); ^{13}C NMR (75 MHz, $CDCl_3$): δ = 170.8, 155.1, 122.3, 77.2, 68.8, 38.6, 34.0, 32.1, 25.3, 18.8, 18.8, 11.4; HRMS-ESI (*m/z*) $[M+Na]^+$ calcd for $C_{18}H_{35}NNaO_2Si$, 348.2329; found: 348.2332.

General Procedure B

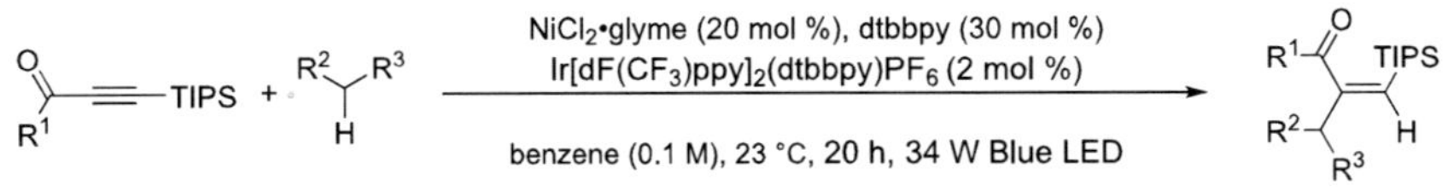

To a vial equipped with a PTFE-coated stirrer bar, $NiCl_2$·glyme (4.39 mg, 0.02 mmol), dtbbpy (8.05 mg, 0.03 mmol), the C(sp^3)–H partner (2.0 mmol), and benzene (1 mL) were added. The resulting solution was stirred for 5 min to give a green suspension before $Ir[dF(CF_3)ppy]_2(dtbbpy)PF_6$ (2.24 mg, 0.002 mmol) and the corresponding alkyne (0.1 mmol) were added to the vial. The resulting mixture was stirred for 20 h under 34 W blue LED irradiation at 23 °C under Ar. The reaction mixture was concentrated *in vacuo* and purified by flash column chromatography (silica gel, hexanes/CH_2Cl_2, gradient elution or hexanes/EtOAc, gradient elution) to afford the desired product.

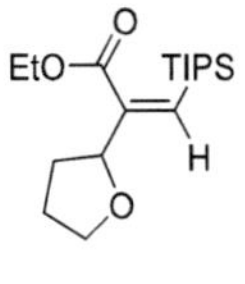

Ethyl (Z)-2-(Tetrahydrofuran-2-yl)-3-(triisopropylsilyl)acrylate (2h)

Yellow oil, 28.3 mg (0.087 mmol, 87% yield); All physical and spectroscopic data is in full agreement with data presented above.

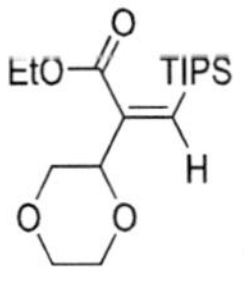

Ethyl (Z)-2-(1,4-Dioxan-2-yl)-3-(triisopropylsilyl)acrylate (2m)
Colorless oil, 25.7 mg (0.075 mmol, 75% yield); ^{1}H NMR (500 MHz, $CDCl_3$): δ = 6.49 (d, J = 1.2 Hz, 1H), 4.49 (ddd, J = 9.5, 2.5, 1.2 Hz, 1H), 4.22 (q, J = 7.1 Hz, 2H), 3.97 (dd, J = 11.3, 2.5 Hz, 1H), 3.87 (dtd, J = 14.5, 11.7, 2.9 Hz, 2H), 3.74 (dd, J = 11.6, 2.6 Hz, 1H), 3.63 (td, J = 11.4, 3.1 Hz, 1H), 3.17 (dd, J = 11.3, 9.6 Hz, 1H), 1.30 (t, J = 7.1 Hz, 3H), 1.26 (dq, J = 14.5, 7.3 Hz, 3H), 1.04 (d, J = 7.4 Hz, 18H); ^{13}C NMR (75 MHz, $CDCl_3$): δ = 166.4, 145.8, 138.9, 76.2, 72.1, 67.1, 66.3, 60.7, 19.0, 19.0, 14.2, 12.1; HRMS-ESI (*m/z*) $[M+Na]^+$ calcd for $C_{18}H_{34}NaO_4Si$, 365.2119; found: 365.2120.

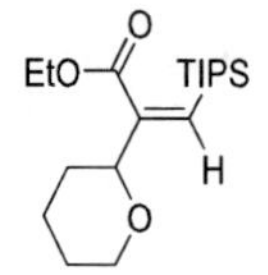

Ethyl (Z)-2-(Tetrahydro-2*H*-pyran-2-yl)-3-(triisopropylsilyl)acrylate (2n)
Colorless oil, 21.8 mg (0.064 mmol, 64% yield); ^{1}H NMR (300 MHz, $CDCl_3$): δ = 6.27 (d, J = 1.3 Hz, 1H), 4.27–4.15 (m, 3H), 4.12–4.04 (m, 1H), 3.54 (td, J = 11.3, 2.9 Hz, 1H), 1.88–1.81 (m, 2H), 1.67–1.49 (m, 3H), 1.34–1.17 (m, 7H), 1.07–1.03 (m, 18H); ^{13}C NMR (75 MHz, $CDCl_3$): δ = 167.7, 151.0, 133.2, 78.2, 68.8, 60.5, 32.5, 25.8, 23.8, 19.1, 19.0, 14.2, 12.2; HRMS-ESI (*m/z*) $[M+Na]^+$ calcd for $C_{19}H_{36}NaO_3Si$, 363.2326; found: 363.2328.

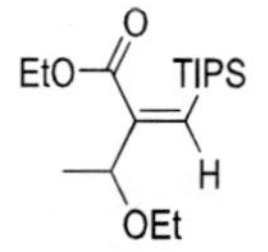

Ethyl (Z)-3-Ethoxy-2-((triisopropylsilyl)methylene)butanoate (2o)
Yellow oil, 10.2 mg (0.031 mmol, 31% yield); ^{1}H NMR (300 MHz, $CDCl_3$): δ = 6.31 (d, J = 0.8 Hz, 1H), 4.33–4.16 (m, 3H), 3.53–3.31 (m, 2H), 1.36–1.18 (m, 12H), 1.05 (dd, J = 7.3, 3.3 Hz, 18H); ^{13}C NMR (75 MHz, $CDCl_3$): δ = 167.9, 151.5, 133.4, 76.5, 64.0, 60.5, 22.7, 19.0, 18.9, 15.4, 14.2, 12.1; HRMS-ESI (*m/z*) $[M+Na]^+$ calcd for $C_{18}H_{36}NaO_3Si$, 351.2326; found: 351.2329.

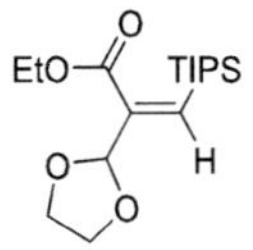

Ethyl (Z)-2-(1,3-Dioxolan-2-yl)-3-(triisopropylsilyl)acrylate (2p)
Colorless oil, 12.8 mg (0.039 mmol, 39% yield); ^{1}H NMR (500 MHz, $CDCl_3$): δ = 6.47 (s, 1H), 5.78 (s, 1H), 4.25 (q, J = 7.1 Hz, 2H), 3.97 (s, 4H), 1.32 (t, J = 7.1 Hz, 3H), 1.31–1.20 (m, 3H), 1.05 (d, J = 7.4 Hz, 18H); ^{13}C NMR (75 MHz, $CDCl_3$): δ = 166.8, 145.8, 137.6, 102.3, 64.8, 60.7, 18.9, 14.2, 12.0; HRMS-ESI (*m/z*) $[M+Na]^+$ calcd for $C_{17}H_{32}NaO_4Si$, 351.1962; found: 351.1962.

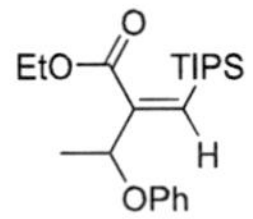

Ethyl (Z)-3-Phenoxy-2-((triisopropylsilyl)methylene)butanoate (2q)
Yellow oil, 17.3 mg (0.046 mmol, 46% yield); ^{1}H NMR (500 MHz, $CDCl_3$): δ = 7.24–7.19 (m, 2H), 6.90–6.84 (m, 3H), 6.32 (d, J = 0.7 Hz, 1H), 5.24 (q, J = 6.4 Hz, 1H), 4.35–4.20 (m, 2H), 1.51 (d, J = 6.3 Hz, 3H), 1.34 (t, J = 7.1 Hz, 3H), 1.18 (dt, J = 14.9, 7.5 Hz, 3H), 0.92 (dd, J = 33.0, 7.5 Hz, 18H); ^{13}C NMR (75 MHz, $CDCl_3$): δ = 167.1, 157.2, 149.3, 136.2, 129.2, 120.7, 115.7, 73.4, 60.8, 22.2, 18.8, 18.8, 14.2, 12.1; HRMS-ESI (*m/z*) $[M+Na]^+$ calcd for $C_{22}H_{36}NaO_3Si$, 399.2326; found: 399.2328.

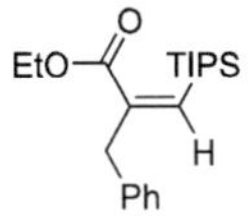

Ethyl (Z)-2-Benzyl-3-(triisopropylsilyl)acrylate (2r)
Colorless oil, 7.6 mg (0.022 mmol, 22% yield); ^{1}H NMR (300 MHz, $CDCl_3$): δ = 7.33–7.14 (m, 5H), 5.99 (d, J = 1.1 Hz, 1H), 4.13 (q, J = 7.1 Hz, 2H), 3.77 (s, 2H), 1.31–1.16 (m, 6H), 1.03 (d, J = 7.3 Hz, 18H); ^{13}C NMR (75 MHz, $CDCl_3$): δ = 167.8, 148.6, 139.2, 139.1, 128.9, 128.3, 126.2, 60.5, 43.6, 19.0, 14.1, 12.2; HRMS-ESI (*m/z*) $[M+Na]^+$ calcd for $C_{21}H_{34}NaO_2Si$ 369.2220; found: 369.2223.

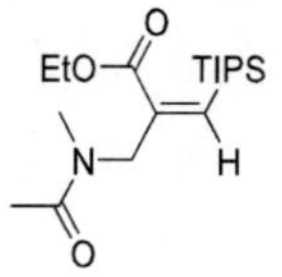

Ethyl (Z)-2-Benzyl-3-(triisopropylsilyl)acrylate (2s) (rotamer, 2.25:1)
Colorless oil, 26.3 mg (0.077 mmol, 77% yield); ^{1}H NMR (300 MHz, $CDCl_3$): δ = 6.00 (t, J = 1.5 Hz, 2.25H), 5.94 (t, J = 1.5 Hz, 1H), 4.34–4.13 (m, 13H), 2.94 (s, 3H), 2.89 (s, 6.75H), 2.13 (s, 3H), 2.06 (s, 6.75H), 1.33–1.22 (m, 19.5H), 1.02 (d, J = 7.3 Hz, 58.5H); ^{13}C NMR (75 MHz, $CDCl_3$): δ = 171.3, 170.4, 167.1, 166.6, 143.4, 143.0, 137.5, 137.1, 61.0, 60.7, 55.0, 51.4, 35.7, 33.5, 21.7, 20.9, 19.0, 18.9, 14.1, 12.1, 12.0; HRMS-ESI (*m/z*) $[M+Na]^+$ calcd for $C_{18}H_{35}NNaO_3Si$, 364.2278; found: 364.2278.

2.4.6 A Representative 1 mmol Scale Reaction

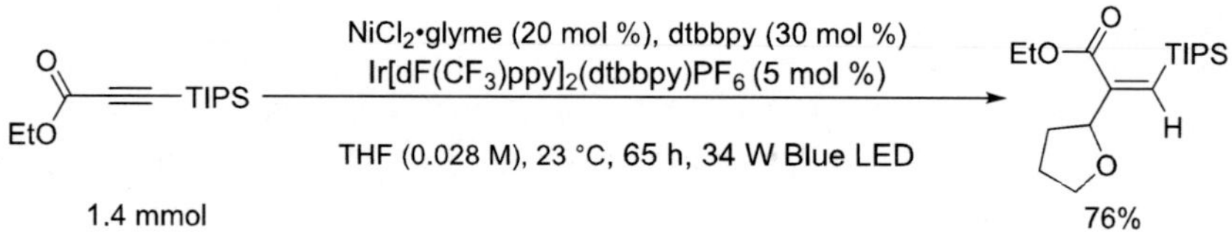

To a vial equipped with a PTFE-coated stirrer bar, $NiCl_2 \cdot$glyme (62 mg, 0.28 mmol), dtbbpy (113 mg, 0.42 mmol), and THF (50 mL) were added. The resulting solution was stirred for 5 min to give a green suspension before $Ir[dF(CF_3)ppy]_2(dtbbpy)PF_6$ (79 mg, 0.07 mmol) and the corresponding alkyne (356 mg, 1.4 mmol) were added to the vial. The resulting mixture was stirred for 65 h under 34 W blue LED irradiation at 23 °C under Ar. Subsequently, the reaction mixture was concentrated *in vacuo* and purified by flash column chromatography (silica gel, hexanes/CH_2Cl_2, gradient elution or hexanes/EtOAc, gradient elution) to afford the desired product (1.06 mmol, 76% yield).

2.4.7 *Post-Functionalization of Enone Products*

Halogenative Desilylation of Alkenyl Silane

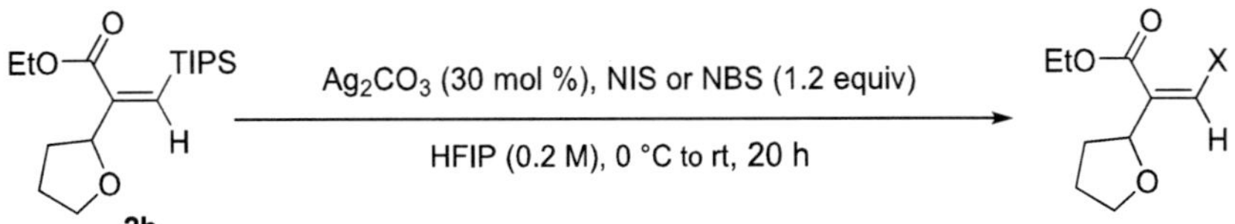

2h

3a (X = I)
3b (X = Br)

To a stirred solution of the corresponding alkenyl silane **2** (39.2 mg, 0.12 mmol) in 1,1,1,3,3,3-hexafluoropropan-2-ol (HFIP, 0.6 mL) at 0 °C, silver carbonate (9.9 mg, 0.036 mmol) was added. The reaction mixture was protected from light before the addition of 1-iodopyrrolidine-2,5-dione (32.4 mg, 0.14 mmol) or 1-bromopyrrolidine-2,5-dione (25.6 mg, 0.14 mmol). When no alkenyl silane was observed on the TLC, the resulting solution was diluted with water (1 mL) and extracted with CH_2Cl_2 (3 × 1 mL). The combined organic extracts were dried (anhydrous Na_2SO_4) and concentrated *in vacuo*. The resulting residue was purified by flash column chromatography (silica gel, hexanes/CH_2Cl_2, gradient elution) to afford the desired product.

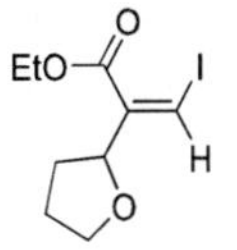

Ethyl (Z)-3-Iodo-2-(tetrahydrofuran-2-yl)acrylate (3a)
Colorless oil, 14.2 mg (0.048 mmol, 40% yield); ^{1}H NMR (300 MHz, $CDCl_3$): δ = 7.13 (d, *J* = 1.6 Hz, 1H), 4.70 (ddd, *J* = 7.4, 6.0, 1.6 Hz, 1H), 4.37–4.23 (m, 2H), 3.99–3.79 (m, 2H), 2.25–2.14 (m, 1H), 1.95–1.85 (m, 2H), 1.84–1.71 (m, 1H), 1.35 (t, *J* = 7.1 Hz, 3H); ^{13}C NMR (75 MHz, $CDCl_3$): δ = 165.7, 144.9, 83.2, 80.0, 68.7, 61.2, 31.8, 25.2, 14.1; HRMS-ESI (*m/z*) $[M+Na]^+$ calcd for $C_9H_{13}INaO_3$, 318.9802; found: 318.9803.

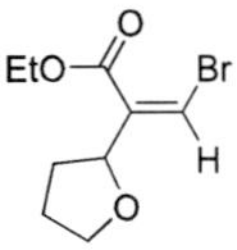

Ethyl (Z)-3-Bromo-2-(tetrahydrofuran-2-yl)acrylate (3b)
Colorless oil, 16.7 mg (0.067 mmol, 56% yield); ^{1}H NMR (500 MHz, $CDCl_3$): δ = 6.81 (d, J = 1.7 Hz, 1H), 4.69 (ddd, J = 7.3, 6.1, 1.6 Hz, 1H), 4.36–4.23 (m, 2H), 3.99–3.90 (m, J = 8.1 Hz, 1H), 3.88–3.80 (m, 1H), 2.26–2.13 (m, 1H), 1.97–1.87 (m, 2H), 1.86–1.73 (m, 1H), 1.35 (t, J = 7.1 Hz, 3H); ^{13}C NMR (75 MHz, $CDCl_3$): δ = 165.3, 139.5, 111.2, 79.0, 68.7, 61.2, 31.6, 25.3, 14.1; HRMS-ESI (*m/z*) $[M+Na]^+$ calcd for $C_9H_{13}BrNaO_3$, 270.9940; found: 270.9944.

Alcohol Synthesis by Carbonyl Reduction

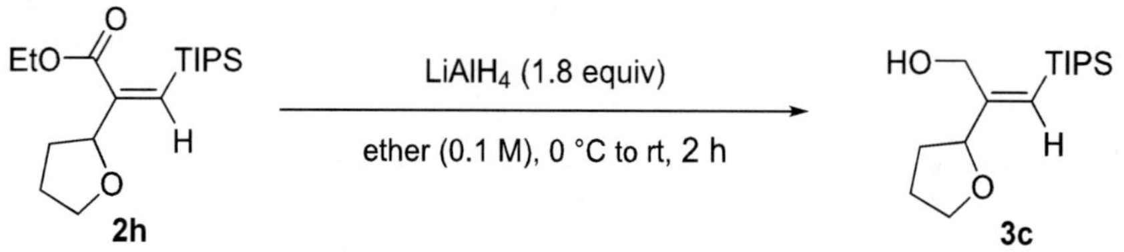

$LiAlH_4$ (8.0 mg, 0.21 mmol) was added to a stirred solution of the corresponding enone **2** (38.5 mg, 0.12 mmol) in ether (1.2 mL) at 0 °C. The resulting mixture was stirred for 1 h at 0 °C. Subsequently, the resulting mixture was warmed to room temperature for 1 h before it was quenched with potassium sodium tartrate tetrahydrate (sat. aq, 2 mL). The layers were separated and the aqueous layer was extracted with ether (3 × 2 mL). The combined organic layers were washed with brine, dried (anhydrous Na_2SO_4), and concentrated *in vacuo*. The resulting residue was purified by flash column chromatography (silica gel, hexanes/EtOAc, gradient elution) to afford **3c** (24.8 mg, 0.087 mmol, 73% yield) as a colorless oil; ^{1}H NMR (500 MHz, $CDCl_3$): δ = 5.50 (d, J = 0.9 Hz, 1H), 4.58–4.50 (m, 1H), 4.28 (d, J = 12.0 Hz, 1H), 4.09 (d, J = 11.9 Hz, 1H), 3.98 (dd, J = 15.1, 7.0 Hz, 1H), 3.82 (dd, J = 14.0, 7.8 Hz, 1H), 2.97 (br, 1H), 2.20–2.07 (m, 1H), 2.01–1.90 (m, 2H), 1.89–1.71 (m, 1H), 1.20–1.01 (m, 21H); ^{13}C NMR (75 MHz, $CDCl_3$): δ = 156.6, 122.8, 85.4, 68.4, 63.8, 32.0, 25.8, 18.8, 12.1; HRMS-ESI (*m/z*) $[M+Na]^+$ calcd for $C_{16}H_{32}NaO_2Si$, 307.2064; found: 307.2063.

Ketone Synthesis Through a Weinreb Amide

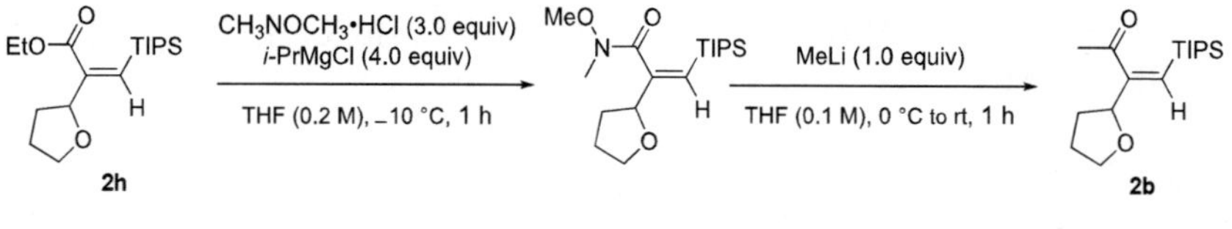

To a stirred suspension of *N,O*-dimethylhydroxylamine hydrochloride (215 mg, 2.2 mmol) with the corresponding enone (240 mg, 0.735 mmol) in THF (3.7 mL) at –30 °C under Ar, ^{i}PrMgCl (2.0 M in THF, 1.5 mL, 2.94 mmol) was added dropwise.

After the addition was completed, the reaction mixture was stirred for 40 min at –10 °C. The resulting mixture was stirred for 1 h before it was quenched with NH_4Cl (sat. aq, 3 mL). The layers were separated and the aqueous layer was extracted with ether (3 × 2 mL). The combined organic extracts were dried (anhydrous Na_2SO_4) and concentrated *in vacuo*. The resulting residue was purified by flash column chromatography (silica gel, hexanes/EtOAc, gradient elution) to afford the corresponding Weinreb amide (157.5 mg, 0.46 mmol, 63% yield) as a yellow oil. Subsequently, to a stirred solution of the corresponding Weinreb amide (157.5 mg, 0.46 mmol) in THF (4.6 mL) at 0 °C under Ar, MeLi (1.6 M in ether, 0.29 mL, 0.46 mmol) was added dropwise. The resulting mixture was stirred for 1 h before quenching with NH_4Cl (sat. aq, 4 mL). The layers were separated and the aqueous layer was extracted with ether (3 × 4 mL). The combined organic layers were dried (anhydrous Na_2SO_4) and concentrated *in vacuo*. The resulting residue was purified by flash column chromatography (silica gel, hexanes/EtOAc, gradient elution) to afford **2b** (85.2 mg, 0.29 mmol, 62% yield). All physical and spectroscopic data is in full agreement with those obtained by direct hydroalkylation of **1b**.

Cross-Coupling of Alkenyl Silane with Arylboronic Acid

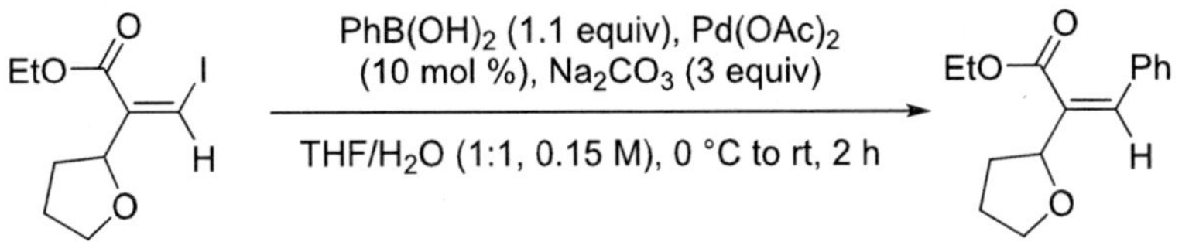

3a **4a**

$Pd(OAc)_2$ (1.5 mg, 0.0068 mmol) was added to a stirred solution of the corresponding alkenyl halide **3a** (20.1 mg, 0.068 mmol), phenylboronic acid (9.1 mg, 0.075), and sodium carbonate (21.6 mg, 0.2 mmol) in THF/H_2O (0.45 mL) at 0 °C. The resulting mixture was stirred for 3 h before quenching with NH_4Cl (sat. aq, 1 mL). The layers were separated and the aqueous layer was extracted with EtOAc (3 × 1 mL). The combined organic layers were dried (anhydrous Na_2SO_4) and concentrated *in vacuo*. The resulting residue was purified by flash column chromatography (silica gel, hexanes/EtOAc, gradient elution) to afford **4a** (13.5 mg, 0.055 mmol, 81% yield) as a colorless oil; ^{1}H NMR (300 MHz, $CDCl_3$): δ = 7.38–7.25 (m, 5H), 6.90 (d, *J* = 1.0 Hz, 1H), 4.79–4.72 (m, 1H), 4.17 (q, *J* = 7.1 Hz, 2H), 4.07–3.98 (m, 1H), 3.95–3.85 (m, 1H), 2.30–2.17 (m, 1H), 2.09–1.92 (m, 3H), 1.14 (t, *J* = 7.1 Hz, 3H); ^{13}C NMR (75 MHz, $CDCl_3$): δ = 168.5, 135.8, 135.7, 131.5, 128.4, 128.1, 127.9, 79.6, 68.7, 60.7, 31.7, 25.6, 13.8; HRMS-ESI (*m/z*) $[M+Na]^+$ calcd for $C_{15}H_{18}NaO_3$, 269.1148; found: 269.1148.

2.4.8 Radical Quenching Study Using TEMPO

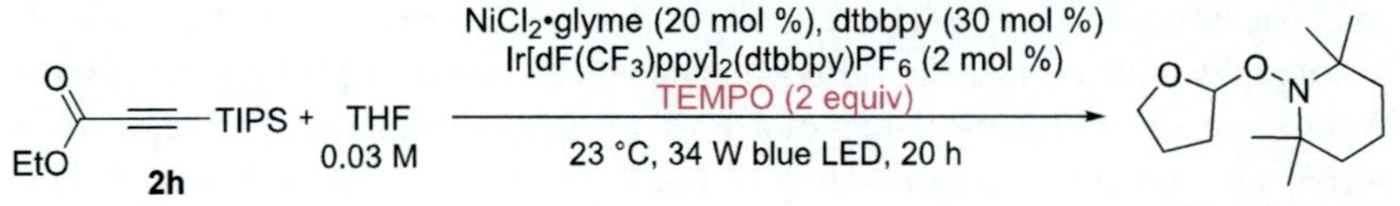

To a vial equipped with a PTFE-coated stirrer bar, $NiCl_2$·glyme (8.79 mg, 0.04 mmol), dtbbpy (16.1 mg, 0.06 mmol), and THF (6 mL) were added. The resulting solution was stirred for 5 min to give a green suspension before Ir[dF(CF_3)ppy]$_2$(dtbbpy)PF_6 (4.49 mg, 0.004 mmol), the corresponding alkyne **1** (50.9 mg, 0.2 mmol), and TEMPO (62.5 mg, 0.4 mmol) were added. The resulting mixture was stirred for 20 h under 34 W blue LED irradiation at 23 °C under Ar. Subsequently, the reaction mixture was concentrated *in vacuo* and analyzed by LC–MS (Fig. 2.8).

A THF-TEMPO adduct was detected indicating the formation of THF radicals by LC–MS and crude NMR. THF-TEMPO: 227.18853 g/mol (exact mass); LC–MS $[M+H]^+$ calcd for $C_{13}H_{26}NO_2$, 228.19581; found: 228.2.

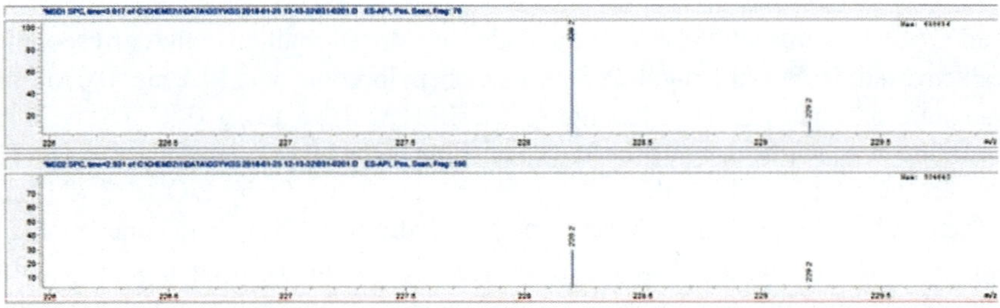

Fig. 2.8 LC–MS data of THF-TEMPO

2.4.9 *Investigation of Regioselectivity Dependence on the Size of Coupling Partner*

See Table 2.6.

2.4.10 *Investigation of a Base Effect*

See Table 2.7.

2.4.11 *Alkenylation Test of THF Radicals by DTBP*

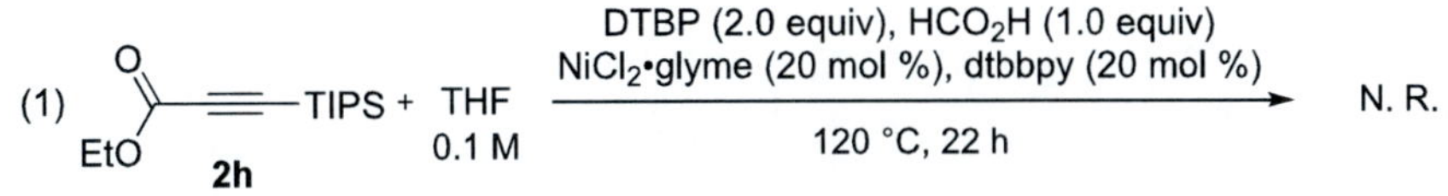

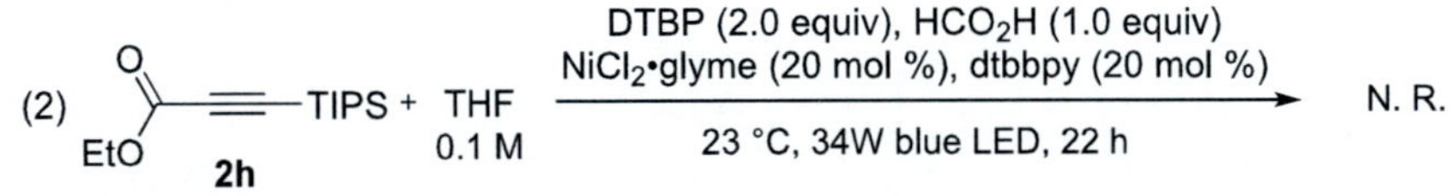

To a stirred solution of alkyne **2h** (50.9 mg, 0.2 mmol) in THF (2 mL), $NiCl_2$·glyme (8.79 mg, 0.04 mmol), dtbbpy (10.74 mg, 0.04 mmol), di-*tert*-butyl peroxide (58.5 mg, 0.4 mmol), and formic acid (9.2 mg, 0.2 mmol) were added. The resulting mixture was stirred for 22 h at 120 °C or under 34 W blue LED irradiation at 23 °C before it was concentrated *in vacuo*. Both the resulting residues were analyzed by ^{1}H NMR and it was evident that the desired products had not formed.

Table 2.6 Observation of regioselectivity dependence on the coupling partner size

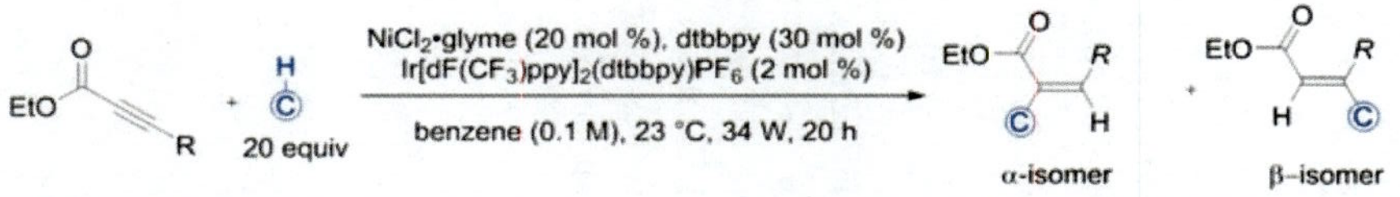

Entry	R	C-H donor	Ni	Ligand	Regioselectivity (α:β)
1	*n*-butyl	*n*-methoxybenzene	20 mol %	30 mol %	2.23:1
2	*n*-butyl	THF	20 mol %	30 mol %	2.88:1
3	*n*-butyl	Diethyl ether	20 mol %	30 mol %	3.39:1
4	*n*-butyl	*n*-ethoxybenzene	20 mol %	30 mol %	4.15:1
5	*n*-butyl	*n*-propoxybenzene	20 mol %	30 mol %	4.20:1
6	*t*-butyl	THF	20 mol %	30 mol %	>10:1
7	TMS	THF	20 mol %	30 mol %	Only α
8	TMS	Diethyl ether	20 mol %	30 mol %	Only α
9	TBS	THF	20 mol %	30 mol %	Only α
10	TBS	Diethyl ether	20 mol %	30 mol %	Only α
11	TBS	Dibutyl ether	20 mol %	30 mol %	Only α
12	TES	THF	20 mol %	30 mol %	Only α
13	TES	Diethyl ether	20 mol %	30 mol %	Only α
14	TIPS	THF	20 mol %	30 mol %	Only α
15	TIPS	Diethyl ether	20 mol %	30 mol %	Only α

Table 2.7 A base-adding experiment

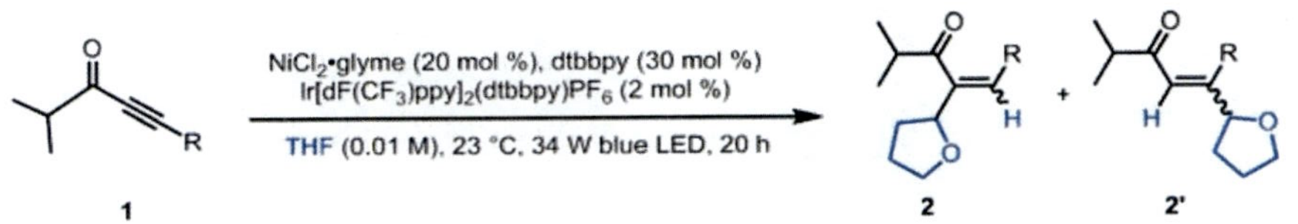

Entry	R	Deviation	Yield (%)	2:2′	*Z*:*E* of 2
1	TIPS	None	83	Only 2	>20:1
3	TIPS	Adding Li_2C0_3(1 equiv)	57	Only 2	>20:1
4	TIPS	Adding Na_2C0_3 (1 equiv)	54	Only 2	>20:1
5	TIPS	Adding K_2C0_3 (1 equiv)	52	Only 2	>20:1
6	TIPS	Adding Cs_2C0_3(1 equiv)	52	Only 2	>20:1
7	TIPS	Adding pyridine (1 equiv)	34	Only 2	>20:1
8	TIPS	TIPS adding DBU (1 equiv)	0	Only 2	>20:1

2.4.12 Deuterium Labeling Study

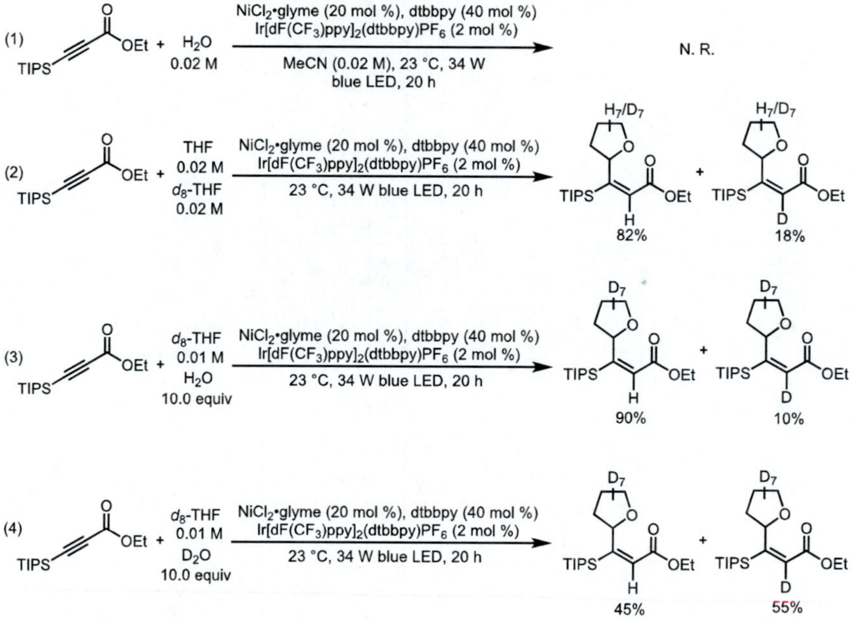

See Figs. 2.9, 2.10, and 2.11.

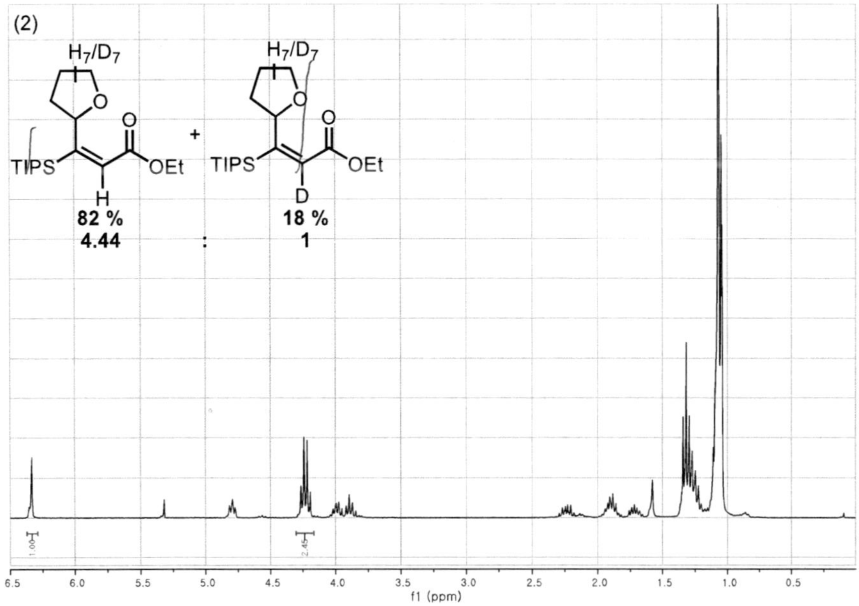

Fig. 2.9 ^{1}H NMR spectrum for observation of the H/D Ratio

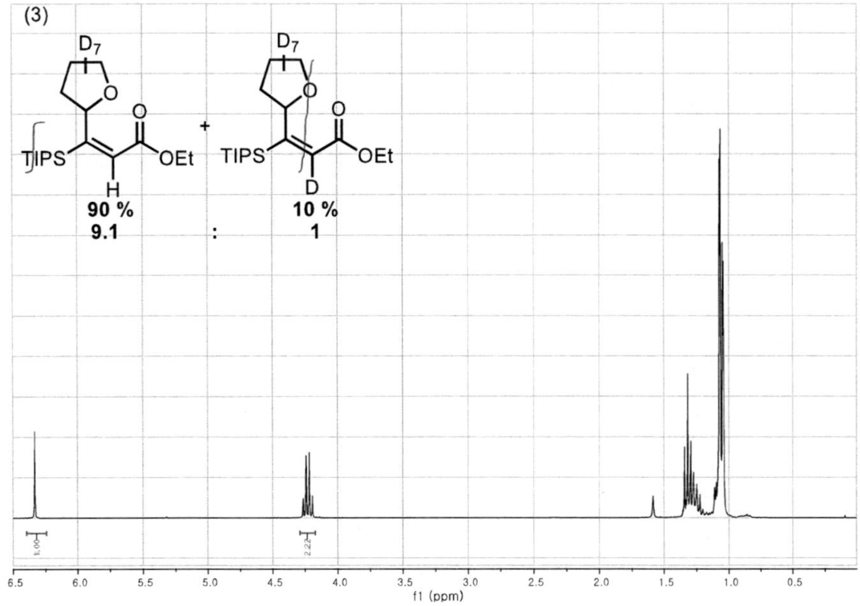

Fig. 2.10 ^{1}H NMR spectrum for observation of the H/D Ratio

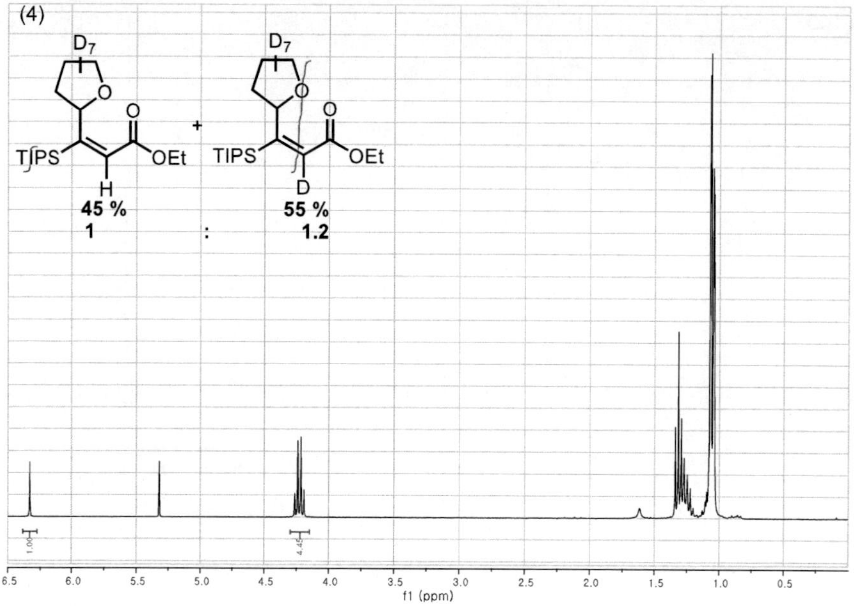

Fig. 2.11 ^{1}H NMR spectrum for observation of the H/D Ratio

References

1. (a) Kabalka GW, Yu S, Li N-S, Lipprandt U (1999) Tetrahedron Lett 40:37. (b) Yu S, Li N-S, Kabalka GW (1999) J Org Chem 64:5822.
2. (a) Wittig G, Schöllkopf U (1954) Chem Ber 87:1318. (b) Maryanoff BE, Reitz AB (1989) Chem Rev 89:863.
3. (a) Simard-Mercier J, Jiang JL, Ho ML, Flynn AB, Ogilvie WW (2008) J Org Chem 73:5899. (b) Dorn SCM, Olsen AK, Kelemen RE, Shrestha R, Weix DJ (2015) Tetrahedron Lett 56:3365.
4. (a) Thibonnet JR, Launay VR, Abarbri M, Duchêne A, Parrain J-L (1998) Tetrahedron Lett 39:4277. (b) Lee J-E, Kwon J, Yun J (2008) Chem Commun 733.
5. (a) Gürtler C, Buchwald SL (1999) Chem Eur J 5:3107. (b) Littke AF, Fu GC (2001) J Am Chem Soc 123:6989.
6. (a) Reddy MC, Jeganmohan M (2013) Chem Commun 49:481. (b) Manikandan R, Jeganmohan M (2015) Org Biomol Chem 13:10420.
7. (a) Yamamoto Y, Kirai N, Harada Y (2008) Chem Commun 2010. (b) Bush AG, Jiang JL, Payne PR, Ogilvie WW (2009) Tetrahedron 65:8502. (c) Hendrix AJM, Jennings MP (2010) Org Lett 12:2750.
8. Chen L, Yang J, Li L, Weng Z, Kang Q (2014) Tetrahedron Lett 55:6096.
9. Punner F, Hilt G (2014) Chem Commun 50:7310.
10. Li J, Zhang J, Tan H, Wang DZ (2015) Org Lett 17:2522.
11. Deng H-P, Fan X-Z, Chen Z-H, Xu Q-H, Wu J (2017) J Am Chem Soc 139:13579.
12. (a) Shields BJ, Doyle AG (2016) J Am Chem Soc 138:12719. (b) Nielsen MK, Shields BJ, Liu J, Williams MJ, Zacuto MJ, Doyle AG (2017) Angew Chem Int Ed 56:7191. (c) Heitz DR, Tellis JC, Molander GA (2016) J Am Chem Soc 138:12715. (d) Till NA, Smith RT, MacMillan DWC (2018) J Am Chem Soc 140:5701. (e) Kang B, Hong SH (2017) Chem Sci 8:6613. (f) Twilton J, Le C, Zhang P, Shaw MH, Evans RW, MacMillan DWC (2017) Nat Rev Chem 1:0052.

13. Ruecker C (1995) Chem Rev 95:1009.
14. Reduced alkenes were the major side products.
15. (a) Ilardi EA, Stivala CE, Zakarian A (2008) Org Lett 10:1727. (b) Sidera M, Costa AM, Vilarrasa J (2011) Org Lett 13:4934.
16. (a) Nahm S, Weinreb SM (1981) Tetrahedron Lett 22:3815. (b) Singh J (2000) J Prakt Chem 342:340. (c) Balasubramaniam S, Aidhen I (2008) Synthesis 3707.
17. Huggins JM, Bergman RG (1981) J Am Chem Soc 103:3002.
18. Lowry MS, Goldsmith JI, Slinker JD, Rohl R, Pascal RA, Malliaras GG, Bernhard S (2005) Chem Mater 17:5712.
19. Edwankar RV, Edwankar CR, Deschamps J, Cook JM (2011) Org Lett 13:5216.
20. Peh G, Floreancig PE (2015) Org Lett 17:3750.
21. Morales-Serna JA, Sauza A, Padrón de Jesús G, Gaviño R, García de la Mora G, Cárdenas J (2013) Tetrahedron Lett 54:7111.
22. Wang Z, Li L, Huang Y (2014) J Am Chem Soc 136:12233.
23. Wang H, Guo LN, Wang S, Duan X-H (2015) Org Lett 17:3054.
24. Hu K, Yang H, Zhang W, Qin Y (2013) Chem Sci 4:3649.
25. Liu X, Yu L, Luo M, Zhu J, Wei W (2015) Chem Eur J 21:8745.
26. Rankin T, Tykwinski RR (2003) Org Lett 5:213.
27. Zhou L, Tang S, Qi X, Lin C, Liu K, Liu C, Lan Y, Lei A (2014) Org Lett 16:3404.

Part II
Development of C(sp^3)–Heteroatom Bond-Forming Reactions via Electrochemical Activation of C(sp^3)–B Bonds and Follow-Up Projects

Chapter 3
Recent Achievements of C(sp^3)–Heteroatom Bond Formation in Electroorganic Synthesis and History of C(sp^3)–B Bond Activation

3.1 Introduction

The backbone structure of organic compounds comprises carbon fragments, while their function is mainly determined by heteroatoms attached to the carbon atom. The diversity of organic molecules containing heteroatoms endows them with novel physical properties that permeate a wide range of research fields in academia and industry [1]. Consequently, the bonding of heteroatoms to carbon atoms has been an enduring subject of investigation for organic chemists.

Over the last decades, C(sp^2)−heteroatom bond-forming reactions have been well developed owing to the emerging new synthetic strategies that generate reactive species [2]. In the case of C(sp^3)−heteroatom bond formation, traditional methods such as Williamson ether synthesis, Mitsunobu reaction, or carbocation chemistry accessed from olefins under strongly acidic conditions have been utilized [3]. However, these methods have remained limited owing to sluggish reactivity and a lack of chemoselectivity. This chapter overviews the electrochemical redox processes and the history of C(sp^3)−B bond activation to design efficient C(sp^3)−heteroatom bond-forming reactions. Representative examples are summarized and discussed based on different substrates involved in the transformation.

S. Y. Go, *Photochemical and Electrochemical Activation Strategies of C(sp3)-Based Building Blocks for Organic Synthesis*, Springer Theses,
https://doi.org/10.1007/978-981-99-8994-2_3

3.2 Advances in the Merger of Electrochemistry and Organic Synthesis

3.2.1 Introduction of Electrosynthesis

Recently, electrochemistry has emerged as an alternative to the utilization of chemical redox agents. The electrochemical strategy enables redox transformations using electric current as the "traceless" reagent and yields environmentally friendly synthetic methods compared to the classical functionalization of organic compounds using dangerous and toxic redox agents. Thus, organic electrochemistry has been considered a useful tool for organic chemists.

Electrosynthesis is classified into direct and indirect (mediated) electrolysis (Fig. 3.1). Direct electrolysis involves heterogeneous electron-transfer between an electrode and a substrate of interest to generate a reactive intermediate. It provides simple reaction conditions and high activation energy while nullifying the need for the compatibility tests of chemical mediators. In contrast, in indirect electrosynthesis, a redox mediator with a lower redox potential than the substrate of interest acts as the electron-transfer-shuttle from the heterogeneous electrode surface to the homogeneous dissolved substrates. Indirect electrolysis is advantageous because (a) the electrolysis can be controlled at lower potentials than the redox potential of the substrate of interest, (b) the reaction rate is relatively fast, and (c) better chemoselectivity can be obtained by avoiding undesirable side reactions. However, the addition of the mediator to the reaction system can interfere with the reaction of the starting materials with reaction coupling partners and disturb the redox process of the reaction substrate [4].

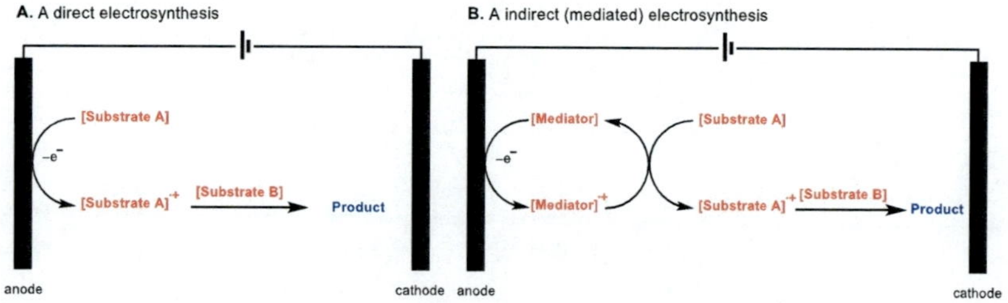

Fig. 3.1 Direct (A) and indirect (mediated) (B) electrosynthesis in the context of anodic oxidation reactions

3.2.2 Representative Electrocatalytic Reactions

Various review articles summarizing the advances in organic electrochemistry have been published [4a]. Pioneering reviews include those reported by Wawzonek and Weinberg in 1968 [5]. Direct oxidation and/or reduction processes were reviewed by Shono [6], Boydston [7], Lei [8], Moeller [9], Schafer [10], Wright [11], and Yoshida [12]. Recent advances in indirect electrosynthesis were described by Francke and Little [13], and more recently by Stahl [14] and Lin [15]. The synthetic applications in complex settings have been described recently by Baran [16]. Bioelectrosynthesis were reviewed by Freguia and Virdis [17] and more recently by Zhu [18] and Minteer [19]. Alternating current electrolysis in organic synthesis was reviewed by Luo [20]. Electrosynthesis in flow chemistry was summarized by Atobe [21], Noel [22], and Pletcher [23]. More recently, specific electrochemical transformation methods have been reviewed extensively, for example, C–H functionalization methods by Ackerman [24], Karkas [25], and Mei [26], heterocycle synthesis by Zeng [27] and Onomura [28], fluorination by Fuchigami [29], dehydrogenative biaryl synthesis by Waldvogel [30], the utilization of *N*-centered radicals by Xu [31], cationic intermediates by Yoshida [32], carboxylic acids by Zhang [33] and Lam [34], and olefin and alkyne functionalization by Ahmed [35], Sun and Han [36], and Lin [37].

Herein, Chapter 3 presents representative achievements of C(sp^3)–heteroatom bond formation in electroorganic synthesis with an emphasis on reaction development and mechanistic insights.

Electrochemical C(sp^3)–H functionalization represents a powerful yet straightforward protocol for constructing carbon–heteroatom bonds (Fig. 3.2). These strategies were investigated by Lei [38]. Single-electron oxidation on the anode causes the direct breaking of N–H bonds to generate *N*-centered radicals. The subsequent 1,5-hydrogen atom transfer (HAT) of the δ–C–H bond by the *N*-centered radical generates a *C*-centered radical. Subsequently, this radical species is oxidized to furnish a carbocation intermediate. Followed by the intramolecular nucleophilic attack of the sulfonamide and subsequent proton elimination, the ring-closure reaction is accomplished.

Intermolecular carbon–heteroatom bond-forming reactions have been revealed through the direct anodic oxidation of carbon–carbon double bonds (Fig. 3.3) [39].

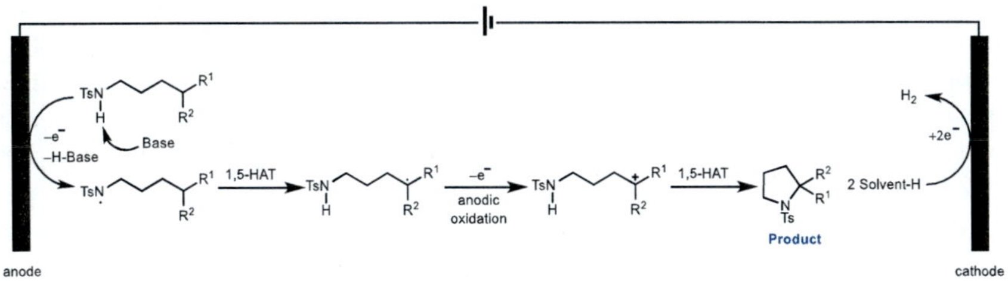

Fig. 3.2 Electrochemical intramolecular C(sp^3)–H amination

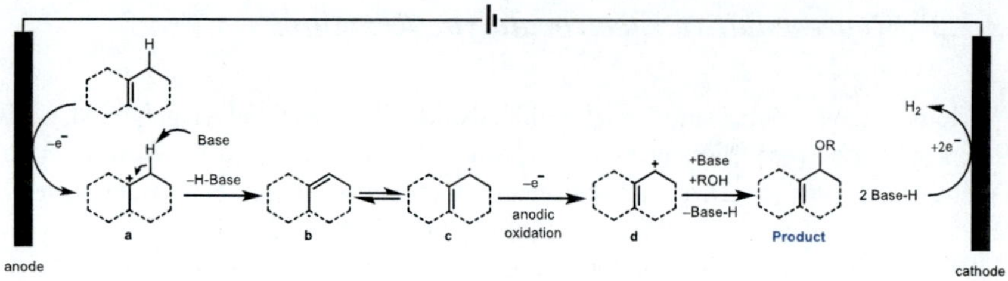

Fig. 3.3 Electrochemical etherification via C(sp^3)–H/O–H cross-coupling

The benzyl or allyl C(sp^3)–H compound is oxidized to the radical cation intermediate (a) via SET. The intermediate generates the alkyl radical (b) through hydrogen abstraction by a base. Subsequently, additional single-electron oxidation of the resonance compound (c) induces the cation intermediate (d) Afterward, a nucleophile reacts with the intermediate (e) to generate the desired product. However, the reactivity scope is limited to etherification products generated from benzyl or allyl C(sp^3)–H compounds owing to the difficulty of direct C(sp^3)–H bond activation.

The synthetically meaningful electrochemical direct C(sp^3)–X bond activation was accomplished by Baran (Fig. 3.4) [40]. The generation of carbocations from unactivated, aliphatic carboxylic acids and their subsequent capture by heteroatom nucleophiles can be leveraged to provide a wide array of hindered carbon–heteroatom bonds. This reaction stems from the oldest electrochemical reaction, the Kolbe dimerization discovered in 1847 [41]. The oxidation of two electrons of a carboxylic acid under mildly alkaline conditions generates a carbocation that can be captured by nucleophiles, which is known as the Hofer–Moest reaction (the so-called interrupted Kolbe variant). The addition of sacrificial supporting oxidants (silver cation) and non-oxidizable bases (2,4,6-collidine) to the reaction system, i.e., Baran's new optimized conditions, considerably expanded the substrate scope, overcoming the challenges of previous S_N2- and carbocation-based approaches in forming sterically hindered, functionalized carbon frameworks. However, even these improved systems suffer from limitations imposed by the difficulties associated with the facile introduction of the carboxyl group to the carbon center and its convenient purification.

To address these challenges, the C(sp^3)–B bond is one of the most convenient functional groups that can be introduced to various positions of organic molecules using borylation methods (Fig. 3.5) [42]. Furthermore, the oxidative liberation of radicals from organoboron compounds has been reported using photoredox-mediated conditions or stoichiometric oxidants [43]. Based on these preliminary studies, electrochemical C(sp^3)–B bond activation was demonstrated by Fuchigami and Stahl [44].

Recently, the Fuchigami group reported that the formation of negatively charged boron-ate complexes decreases the oxidation potentials of C(sp^3)–B bonds. Further, they demonstrated the synthetic availability of organoboron compounds under electrochemical conditions (Fig. 3.6A). However, this method was applicable to the

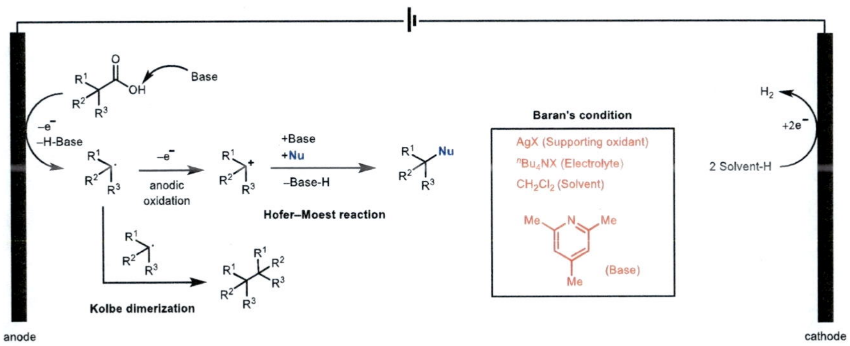

Fig. 3.4 Kolbe dimerization and Hofer–Moest reaction

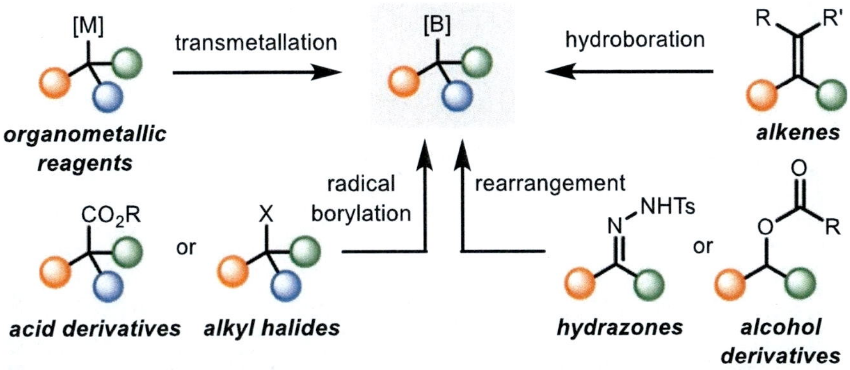

Fig. 3.5 Representative synthetic methods of alkylboron compounds

reaction of primary alkylboron compounds containing α-heteroatom or β-aryl group with solvent quantities of heteroatom-based nucleophiles. Meanwhile, the Stahl group presented that radical reactivity can be controlled using ferrocene-based electron transfer mediators. In other words, the radicals generated from alkylboron compounds using mediators can be utilized for productive intermolecular reactions (Fig. 3.6B). However, the synthetic feasibility of electrochemical $C(sp^3)–B$ bond functionalization has been underexplored, although the basic reactivity of $C(sp^3)–B$ bonds has been reported.

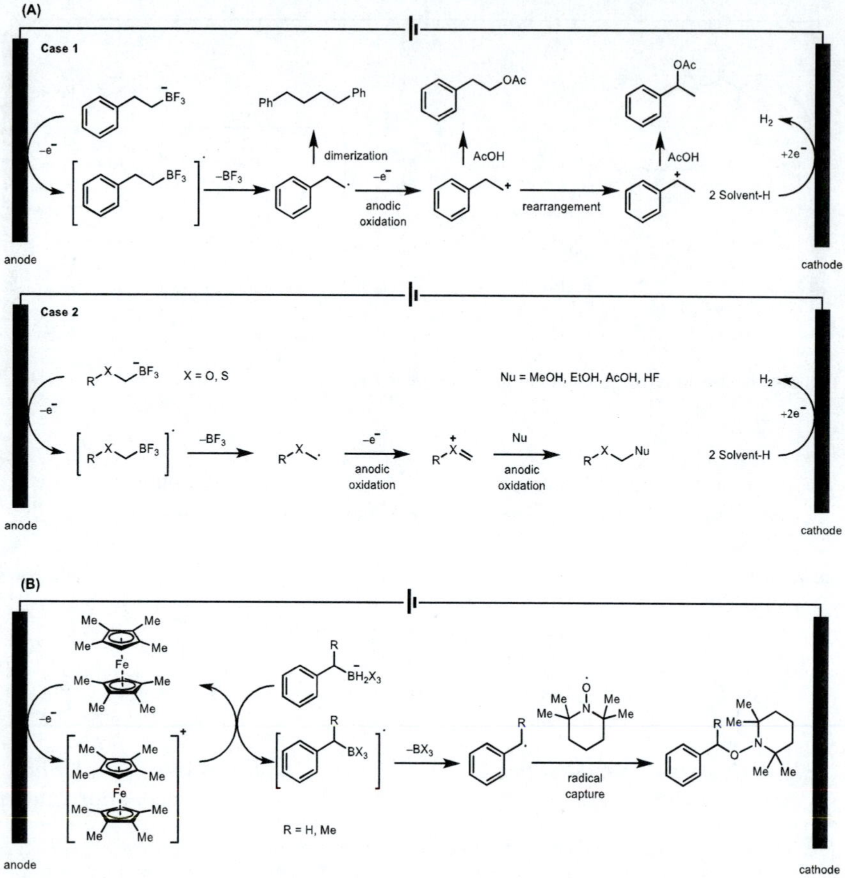

Fig. 3.6 Electrochemical functionalization of organoboron reagents

3.3 History of C(sp^3)–B Bond Activation

Since Herbert Brown developed a C(sp^3)–B bond-forming reaction in the late 1950s [45], organoboron compounds are one of the most developed and applied reagents in organic chemistry (Fig. 3.7). The classical activation of alkylboron reagents is established based on the 1,2-metallate rearrangement of their ate complexes. This migration of the C(sp^3)–B σ bond to the adjacent atom (carbon, nitrogen, and oxygen) upon the loss of a leaving group or two-electron transfer to another atom enables the formation of diverse C–X bonds (X = C, N, and O) [45, 46].

Gradually, the carbanionic nature induced by the electronegativity difference between carbon and boron has been utilized for the development of intermolecular reactions involving electrophiles. The Lewis acid-promoted additions of alkylboron reagents to aldehydes emerged at an early stage, followed by the development of transition-metal-catalyzed coupling reactions called Suzuki–Miyaura coupling [47].

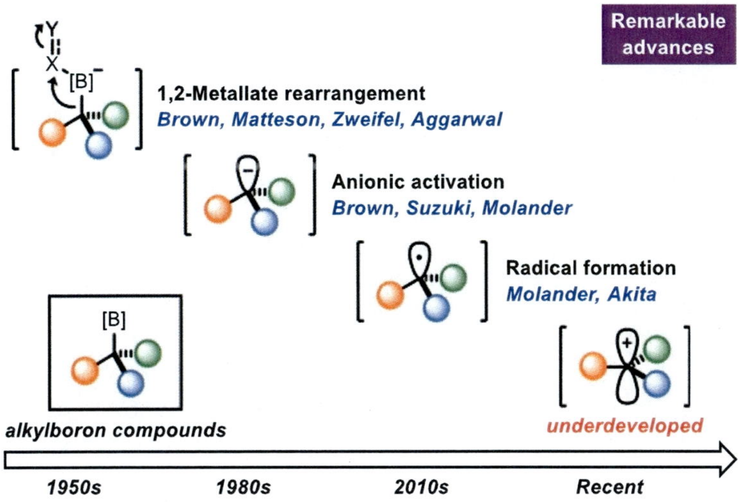

Fig. 3.7 Activation strategies of alkylboron compounds

However, the slow transmetalation rate of alkylboron reagents, the rate-limiting step of Suzuki–Miyaura coupling, limits the availability of boron chemistry compared to the efficient reactivity of arylboron reagents. Further, two-electron transmetalation requires a stoichiometric base and high temperature.

Most recently, SET-based approaches have extended the reactivity scope of $C(sp^3)$–B bonds by generating radical intermediates from alkylboron compounds. Furthermore, the merger of photoredox and transition-metal catalysis has considerably advanced the availability of boron chemistry by enabling the radical activation of $C(sp^3)$–B bonds under mild conditions, nullifying the need for a stoichiometric base or high temperature [48]. Meanwhile, the carbocationic activation of alkylboron compounds despite its suitability for interactions with polar heteroatoms has not been well explored in the synthetic context because of the constraints enforced by the reaction parameters; for example, the unfavorable stoichiometry and/or harsh conditions. Furthermore, a systematic investigation of the reaction mechanism has not been conducted to examine the generation of the carbocation.

3.4 Conclusion

Electrochemistry has been recognized as an efficient alternative to the utilization of chemical redox agents. It enables redox transformations using an electric current as the "traceless" reagent and yields in environmentally friendly synthetic methods. The merger of electrochemistry and organic synthesis has expanded the reactivity scope and improved the selectivity and reaction conditions in many challenging transformations. However, challenges remain in electrochemical $C(sp^3)$–heteroatom bond-forming methodologies, especially in the context of practical synthetic platform design and the installation of suitable electroauxilaries.

Further research directions should be focused on resolving the current limitations of this field. Currently, one of the optimal electroauxilaries that can address these synthetic challenges is the carbon−boron bond. Only limited examples demonstrated the possibility to use C(sp^3)−B bonds for electroorganic synthesis, and a handful of reports presented the explanation of the reaction mechanism to investigate the generation of the carbocation from alkylboron compounds. Therefore, synthetic platforms must be developed and in-depth mechanistic studies must be conducted for the efficient functionalization of organoboron compounds.

References

1. (a) Hartwig JF (2008) Carbon−Heteroatom bond formation catalysed by organometallic complexes. Nature 455:314−322. For examples of important functional molecules with heteroatom substituents: (b) Herzon SB (2017) The mechanism of action of (−)-Lomaiviticin A. Acc Chem Res 50:2577−2588. (c) Sather AC, Lee HG, De La Rosa VY, Yang Y, Müller P, Buchwald SL (2015) A fluorinated Ligand enables room-temperature and Regioselective Pd-catalyzed fluorination of Aryl Triflates and Bromides. J Am Chem Soc 137:13433−13438. (d) Chen CH, Tang CW (2001) Efficient Green Organic Light-Emitting Diodes with Stericly Hindered Coumarin Dopants. Appl Phys Lett 79:3711−3713. (e) Martin SJ, Bradley DDC, Lane PA, Mellor H, Burn PL (1999) Linear and nonlinear optical properties of the conjugated polymers PPV and MEH-PPV. Phys Rev B 59:15133−15142
2. For review, see: (a) Ruiz-Castillo P, Buchwald SL (2016) Applications of Palladium-catalyzed C−N Cross-coupling reactions. Chem Rev 116:12564−12649. (b) Chen J-Q, Li J-H, Dong Z-B (2020) A review on the latest progress of Chan-Lam coupling reaction. Adv Synth Catal 362:3311−3331. (c) Barcellos AM, Sacramento M, da Costa GP, Perin G, João Lenardão E, Alves D (2021) Organoboron compounds as versatile reagents in the transition metal-catalyzed C−S, C−Se and C−Te Bond Formation. Coord Chem Rev 442:214012. (d) Cavedon C, Seeberger PH, Pieber B (2020) Photochemical strategies for Carbon−Heteroatom bond formation. Eur J Org Chem 2020:1379−1392. (e) Zhu C, Yue H, Jia J, Rueping M (2021) Nickel-catalyzed C−Heteroatom cross-coupling reactions under mild conditions via facilitated reductive elimination. Angew Chem, Int Ed 60:17810−17831. For a recently reported electrochemical apporoach, see: (f) Walker BR, Manabe S, Brusoe AT, Sevov CS (2021) Mediator-enabled electrocatalysis with Ligandless Copper for anaerobic Chan-Lam coupling reactions. J Am Chem Soc 143:6257−6265
3. (a) Williamson W (1851) Ueber die theorie der aetherbildung. Justus Liebigs Ann Chem 77, 37−49. (b) Swamy KCK, Kumar NNB, Balaraman E, Kumar KVPP (2009) Mitsunobu and related reactions: advances and applications. Chem Rev 109:2551−2651. (c) Fletcher S (2015) The Mitsunobu Reaction in the 21st Century. Org Chem Front 2:739−752. (d) Beyerman HC, Heiszwolf GJ (1965) Reaction of Steroidal Alcohols with Isobutene. Usefulness of *t*-Butyl as a Hydroxyl-protecting group in a synthesis of Testosterone. Recl Trav Chim Pays-Bas 84:203–212. (e) Smith MB, March J (2007) March's Advanced Organic Chemistry. Wiley, pp 1037–1041
4. (a) Malapit CA, Prater MB, Cabrera-Pardo JR, Li M, Pham TD, McFadden TP, Blank S, Minteer SD (2022) Advances on the merger of Electrochemistry and transition metal catalysis for organic synthesis. Chem Rev 122:3180–3218. (b) Zhu C, Ang NWJ, Meyer TH, Qiu Y, Ackermann L (2021) Organic Electrochemistry: molecular syntheses with potential. ACS Cent Sci 7:415–431

5. Weinberg NL, Weinberg HR (1968) Electrochemical oxidation of organic compounds. Chem Rev 6:449–523
6. Shono T (1984) Electroorganic Chemistry in organic synthesis. Tetrahedron 40:811–850
7. Ogawa KA, Boydston AJ (2015) Recent developments in organocatalyzed Electroorganic Chemistry. Chem Lett 44:10–16
8. Yuan Y, Lei A (2019) Electrochemical Oxidative cross-coupling with hydrogen evolution reactions. Acc Chem Res 52:3309–3324
9. (a) Moeller KD (2017) Intramolecular Carbon–Carbon bond forming reactions at the Anode. In: Steckhan E (ed) Electrochemistry VI Electroorganic synthesis: bond formation at Anode and Cathode. Springer, Berlin, Heidelberg, 1997, pp 49−86. (b) Feng R, Smith JA, Moeller KD (2017) Anodic cyclization reactions and the mechanistic strategies that enable optimization. Acc Chem Res 50:2346−2352
10. Schäfer HJ (2014) Carbon−Carbon bond formation via electron transfer: Anodic coupling. ChemCatChem 6:2792−2795
11. Sperry JB, Wright DL (2006) The application of Cathodic reductions and Anodic oxidations in the synthesis of complex molecules. Chem Soc Rev 35, 605−621
12. Yoshida J, Kataoka K, Horcajada R, Nagaki A (2008) Modern strategies in electroorganic synthesis. Chem Rev 108:2265−2299
13. Francke R, Little RD (2014) Redox catalysis in organic electrosynthesis: basic principles and recent developments. Chem Soc Rev 43:2492−2521
14. Wang F, Stahl SS (2020) Electrochemical oxidation of organic molecules at lower overpotential: accessing broader functional group compatibility with electron-proton transfer mediators. Acc Chem Res 53:561−574
15. Novaes LFT, Liu J, Shen Y, Lu L, Meinhardt JM, Lin S (2021) Electrocatalysis as an enabling technology for organic synthesis. Chem Soc Rev 50:7941−8002
16. (a) Horn EJ, Rosen BR, Baran PS (2016) Synthetic organic electrochemistry: an enabling and innately sustainable method. ACS Cent Sci 2:302−308. (b) Yan M, Kawamata Y, Baran PS (2017) Synthetic organic electrochemical methods since 2000: on the verge of a renaissance. Chem Rev 117:13230−13319
17. Freguia S, Virdis B, Harnisch F, Keller J (2012) Bioelectrochemical systems: microbial versus enzymatic catalysis. Electrochim Acta 82:165−174
18. Wu R, Ma C, Zhu Z (2020) Enzymatic electrosynthesis as an emerging electrochemical synthesis platform. Curr Opin Electrochem 19:1−7
19. (a) Hickey DP, Milton RD, Rasmussen M, Abdellaoui S, Nguyen K, Minteer SD (2015) Fundamentals and applications of bioelectrocatalysis. Electrochemistry; The Royal Society of Chemistry. Cambridge, Vol 13, pp 97−132. (b) Chen H, Simoska O, Lim K, Grattieri M, Yuan M, Dong F, Lee YS, Beaver K, Weliwatte S, Gaffney EM, Minteer SD (2020) Fundamentals, applications, and future directions of bioelectrocatalysis. Chem Rev 120:12903−12993
20. Rodrigo S, Gunasekera D, Mahajan JP, Luo L (2021) Alternating current electrolysis for organic synthesis. Curr Opin Electrochem 28:100712
21. Atobe M, Tateno H, Matsumura Y (2018) Applications of flow microreactors in electrosynthetic processes. Chem Rev 118:4541−4572
22. Noel T, Cao Y, Laudadio G (2019) The fundamentals behind the use of flow reactors in electrochemistry. Acc Chem Res 52:2858−2869
23. Pletcher D, Green RA, Brown RCD (2018) Flow electrolysis cells for the synthetic organic chemistry laboratory. Chem Rev 118:4573−4591
24. Ackermann L (2020) Metalla-electrocatalyzed C−H activation by earth-abundant 3d metals and beyond. Acc Chem Res 53:84−104
25. Karkas MD (2018) Electrochemical strategies for C−H functionalization and C−N bond formation. Chem Soc Rev 47:5786−5865
26. Jiao KJ, Xing YK, Yang QL, Qiu H, Mei TS (2020) Site selective C−H functionalization via synergistic use of electrochemistry and transition metal catalysis. Acc Chem Res 53:300−310
27. Jiang Y, Xu K, Zeng C (2018) Use of electrochemistry in the synthesis of heterocyclic structures. Chem Rev 118:4485−4540

28. Yamamoto K, Kuriyama M, Onomura O (2020) Anodic oxidation for the stereoselective synthesis of heterocycles. Acc Chem Res 53:105–120
29. Fuchigami T, Inagi S (2020) Recent advances in electrochemical systems for selective fluorination of organic compounds. Acc Chem Res 53:322–334
30. Rockl JL, Pollok D, Franke R, Waldvogel SR (2020) A decade of electrochemical dehydrogenative C,C-coupling of Aryls. Acc Chem Res 53:45–61
31. Xiong P, Xu HC (2019) Chemistry with electrochemically generated *N*-centered radicals. Acc Chem Res 52:3339–3350
32. Yoshida JI, Shimizu A, Hayashi R (2018) Electrogenerated Cationic reactive intermediates: the pool method and further advances. Chem Rev 118:4702–4730
33. Chen N, Ye Z, Zhang F (2021) Recent progress on electrochemical synthesis involving carboxylic acids. Org Biomol Chem 19:5501–5520
34. Leech MC, Lam K (2020) Electrosynthesis using carboxylic acid derivatives: new tricks for old reactions. Acc Chem Res 53:121–134
35. (a) Martins GM, Shirinfar B, Hardwick T, Ahmed N (2019) A green approach: vicinal oxidative electrochemical Alkene difunctionalization. ChemElectroChem 6:1300–1315. (b) Martins GM, Shirinfar B, Hardwick T, Murtaza A, Ahmed, N (2019) Organic electrosynthesis: electrochemical Alkyne functionalization. Catal Sci Technol 9:5868–5881
36. Mei H, Yin Z, Liu J, Sun H, Han J (2019) Recent advances on the electrochemical difunctionalization of Alkenes/Alkynes. Chin J Chem 37:292–301
37. (a) Sauer GS, Lin S (2018) An electrocatalytic approach to the radical difunctionalization of Alkenes. ACS Catal 8:5175–5187. (b) Siu JC, Fu N, Lin S (2020) Catalyzing electrosynthesis: a homogeneous electrocatalytic approach to reaction discovery. Acc Chem Res 53:547–560
38. Hu X, Zhang G, Bu F, Nie L, Lei A (2018) Electrochemical-oxidation-induced site-selective intramolecular C(sp^3)–H amination. ACS Catal 8:9370–9375
39. Wang H, Liang K, Xiong W, Samanta S, Li W, Lei A (2020) Electrochemical oxidation-induced etherification via C(sp^3)–H/O–H cross-coupling. Sci Adv 6:eaaz0590
40. (a) Xiang J, Shang M, Kawamata Y, Lundberg H, Reisberg SH, Chen M, Mykhailiuk P, Beutner G, Collins MR, Davies A, Bel MD, Gallego GM, Spangler JE, Starr J, Yang S, Blackmond DG, Baran PS (2019) Hindered Dialkyl Ether synthesis with electrogenerated carbocations. Nature 573:398–402. (b) Sheng T, Zhang H-J, Shang M, He C, Vantourout JC, Baran PS (2020) Electrochemical Decarboxylative *N*-Alkylation of heterocycles. Org Lett 22:7594–7598
41. Kolbe H (1847) Beobachtungen über die oxydirende wirkung des sauerstofs, wenn derselbe mit hülfe einer elektrischen säule entwickelt wird. J Prakt Chem 41:137–139
42. For review, see: (a) Leonori D, Aggarwal VK (2014) Lithiation–Borylation methodology and its application in synthesis. Acc Chem Res 47:3174–3183. (b) Fyfe JWB, Watson AJB (2017) Recent developments in Organoboron Chemistry: old dogs, new tricks. Chem 3:31–55. (c) Friese FW, Studer A (2019) New avenues for C-B bond formation via radical intermediates. Chem Sci 10:8503–8518. (d) Wang M, Shi Z (2020) Methodologies and strategies for selective Borylation of C–Het and C–C bonds. Chem Rev 120:7348–7398. (e) Das KK, Paul S, Panda S (2020) Transition metal-free synthesis of Alkyl Pinacol Boronates. Org Biomol Chem 18:8939–8974. (f) Das KK, Panda S (2020) Functionalization of heterocycles through 1,2-Metallate rearrangement of Boronate complexes. Chem Eur J 26:14270–14282. (g) Manna S, Das KK, Nandy S, Aich D, Paul S, Panda S (2021) A new avenue for the preparation of organoboron compounds via Nickel catalysis. Coord Chem Rev 448:214165. (h) Lai D, Ghosh S, Hajra A (2021) Light-Induced Borylation: Developments and Mechanistic Insights. Org Biomol Chem 19:4397–4428. (i) Kanti Das K, Manna S, Panda S (2021) Transition metal catalyzed asymmetric multicomponent reactions of unsaturated compounds using organoboron reagents. Chem Commun 57:441–459. Representatively, the following protocols were utilized to prepare the alkylboron substrates used in this study (j) Clary JW, Rettenmaier TJ, Snelling R, Bryks W, Banwell J, Wipke WT, Singaram B (2011) Hydride as a leaving group in the reaction of Pinacolborane with Halides under ambient Grignard and Barbier conditions. One-Pot synthesis of Alkyl, Aryl, Heteroaryl, Vinyl, and Allyl Pinacolboronic Esters. J Org Chem 76:9602–9610. (k) Roesner S, Brown CA, Mohiti M, Pulis AP, Rasappan R, Blair DJ, Essafi S, Leonori D,

Aggarwal VK (2014) Stereospecific conversion of alcohols into Pinacol Boronic Esters using Lithiation–Borylation methodology with Pinacolborane. Chem Commun 50:4053–4055. (l) Atack TC, Cook SP (2016) Manganese-catalyzed Borylation of unactivated Alkyl Chlorides. J Am Chem Soc 138:6139–6142. (m) Li C, Wang J, Barton LM, Yu S, Tian M, Peters DS, Kumar M, Yu AW, Johnson KA, Chatterjee AK, Yan M, Baran PS (2017) Decarboxylative Borylation. Science 356:eaam7355. (n) Yang Y, Tsien J, Ben David A, Hughes JME, Merchant RR, Qin T (2021) Practical and modular construction of C(sp^3)–Rich Alkylboron compounds. J Am Chem Soc 143:471–480. (o) Yang Y, Tsien J, Hughes JME, Peters BK, Merchant RR, Qin T (2021) An intramolecular coupling approach to Alkyl Bioisosteres for the synthesis of multisubstituted Bicycloalkylboronates. Nat Chem 13:950–955

43. (a) Yasu Y, Koike T, Akita M (2012) Visible light-induced selective generation of radicals from Organoborates by Photoredox catalysis. Adv Synth Catal 354:3414–3420. (b) Tellis JC, Primer DN, Molander GA (2014) Dual catalysis. Single-Electron transmetalation in Organoboron cross-coupling by Photoredox/Nickel dual catalysis. Science 345:433–436. (c) Seiple IB, Su S, Rodriguez RA, Gianatassio R, Fujiwara Y, Sobel AL, Baran PS (2010) Direct C–H Arylation of electron-deficient heterocycles with Arylboronic acids. J Am Chem Soc 132:13194–13196. (d) Sorin G, Mallorquin RM, Contie Y, Baralle A, Malacria M, Goddard JP, Fensterbank L (2010) Oxidation of Alkyl Trifluoroborates: An opportunity for tin-free radical chemistry. Angew Chem, Int Ed 49:8721–8723
44. (a) Suzuki J, Tanigawa M, Inagi S, Fuchigami T (2016) Electrochemical oxidation of organotrifluoroborate compounds. ChemElectroChem 3:2078–2083. (b) Tanigawa M, Kuriyama Y, Inagi S, Fuchigami T (2016) Electrochemical properties and reactions of Sulfur-containing Organoboron compounds. Electrochim Acta 199:314–318. (c) Inagi S, Fuchigami T (2017) Electrochemical properties and reactions of Organoboron compounds. Curr Opin Electrochem 2:32–37. (d) Ohtsuka K, Inagi S, Fuchigami T (2017) Electrochemical properties and reactions of oxygen-containing Organotrifluoroborates and their Boronic Acid Esters. ChemElectroChem 4:183–187. (e) Lennox AJJ, Nutting JE, Stahl SS (2018) Selective electrochemical generation of Benzylic Radicals enabled by Ferrocene-based Electron-transfer mediators. Chem Sci 9:356–361
45. Brown HC, Rao BCS (1956) A new technique for the conversion of Olefins into Organoboranes and related alcohols. J Am Chem Soc 78:5694–5695
46. (a) Matteson DS, Mah RWH (1963) Neighboring Boron in nucleophilic displacement. J Am Chem Soc 85:2599–2603. (b) Zweifel G, Arzoumanian H, Whitney CC (1967) A convenient stereoselective synthesis of substituted Alkenes via Hydroboration–Iodination of Alkynes. J Am Chem Soc 89:3652–3653. (c) Aggarwal VK, Fang GY, Schmidt AT (2005) Synthesis and applications of Chiral Organoboranes generated from Sulfonium Ylides. J Am Chem Soc 127:1642–1643
47. (a) Brown HC, Jadhav PK (1983) Asymmetric Carbon–Carbon bond formation via *B*-Allyldiisopinocampheylborane. Simple synthesis of secondary Homoallylic alcohols with excellent enantiomeric purities. J Am Chem Soc 105:2092–2093. (b) Miyaura N, Ishiyama T, Ishikawa M, Suzuki A (1986) Palladium-catalyzed cross-coupling reactions of B-Alkyl-9-BBN or Trialkylboranes with Aryl and 1-Alkenyl Halides. Tetrahedron Lett 27:6369–6372. (c) Molander GA, Ito T (2001) Cross-coupling reactions of Potassium Alkyltrifluoroborates with Aryl and 1-Alkenyl Trifluoromethanesulfonates. Org Lett 3:393–396
48. (a) Molander GA, Colombel V, Braz VA (2011) Direct Alkylation of Heteroaryls using Potassium Alkyl- and Alkoxymethyltrifluoroborates. Org Lett 13:1852–1855. (b) Yasu Y, Koike T, Akita M (2012) Visible light-induced selective generation of radicals from Organoborates by Photoredox catalysis. Adv Synth Catal 354:3414–3420. (c) Tellis JC, Primer DN, Molander GA (2014) Single-electron transmetalation in Organoboron cross-coupling by Photoredox/Nickel dual catalysis. Science 345:433. (d) Duret G, Quinlan R, Bisseret P, Blanchard N (2015) Boron Chemistry in a new light. Chem Sci 6:5366–5382

Chapter 4
Introduction of Heteroatoms to Alkyl Carbocations Generated from Alkylboron Reagents via Electrochemical Activation

4.1 Introduction

Organic molecules consist of hydrocarbon frameworks and their function mainly depends on the presence of polar heteroatoms that are attached to the carbon atom. The diversity given to the organic molecules containing heteroatoms creates unique physical properties, which expands the organic chemical space into a range of research areas in academia and industry [1]. During the last decades, C(sp^2)–heteroatom bond formation has dramatically been advanced, as emerging synthetic strategies enable the utilization of more reactive species [2]. On the other hand, biomolecular nucleophilic substitution (S_N2) strategies, such as facilitated displacement or in situ activation of a leaving group, have been still utilized for the formation of C(sp^3)–heteroatom bonds [3]. However, these S_N2 approaches are extremely hampered by the steric congestion at a given reaction center. Recently, in order to overcome numerous problems in S_N2 reaction mechanism, remarkable progress has been made through the generation of highly reactive intermediates [4]. Still, even these improved systems suffer from limitations related to the practicality of the functional handle such as facile introduction, convenient purification [4a–e], and/or chemoselective activation [4f–i].

The use of C(sp^3)–B bonds could be one of the good options that can address these synthetic challenges. The borylation of alkyl compounds has been well developed and dramatically advanced, furnishing the chance that the most convenient functional handle can be introduced to various positions on organic molecules (Fig. 4.1A) [5]. In this context, C(sp^3)–B bond activation strategies have also been established in synthetically useful manners (Fig. 4.1B). The long-established activation strategies of alkylboron compounds have evolved on the basis of the 1,2-rearrangement of their ate complex [6]. This migration of the C(sp^3)–B bond to the adjacent atom (carbon, nitrogen, and oxygen) with loss of a leaving group or two-electron transfer to another atom provides diverse C–X bond formation (X = C, N and O). Over time, the carbanionic character of alkylboron compounds has been extensively utilized as

S. Y. Go, *Photochemical and Electrochemical Activation Strategies of C(sp3)-Based Building Blocks for Organic Synthesis*, Springer Theses,
https://doi.org/10.1007/978-981-99-8994-2_4

electrophilic reaction counterparts [7]. Recently, the single-electron transfer (SET) leading a radical intermediate from boron-containing compounds has widened the reactivity scope by providing a tool that can mediate numerous transformations [8]. On the other hand, two-electron oxidative approaches to generate a carbocation from an alkylboron compound, despite its suitability for interacting with polar heteroatoms, are extremely rare in view of the practical organic synthesis. So far, the basic reactivity has been demonstrated that alkylboron species could be utilized to generate carbocation intermediates under oxidative conditions in both chemical and electrochemical manners [9]. However, attempts to explore the synthetic feasibility of this strategy are still in their infancy due to the constraints such as unfavorable stoichiometry and/or harsh conditions. Moreover, there is no systematic investigation of the activation process to elucidate the formation of the carbocation intermediates.

We suggested that the carbocationic activation of the alkylboron reagent could realize a universally applicable method for the introduction of heteroatoms from group 14 to group 17 (Fig. 4.1C). Of the numerous oxidation strategies available, electrochemistry enables redox transformations using an electric current as the "traceless" reagent compared to classical activation of organic compounds using dangerous and toxic redox agents [10]. Among various organoboron reagents, organotrifluoroborate salts were considered optimal reaction precursors because of (a) their stability for a practical synthetic platform [11] and (b) their anodically active nature for an efficient electroauxiliary [9c–f]. Particularly, they have been known as suitable candidates for electrochemical activation, as demonstrated in the studies of Fuchigami and Stahl [8e, 9c–f, 12]. In this context, we anticipate that the oxidative liberation of intermediates from organoboron compounds should furnish a powerful synthetic platform to solve important synthetic problems with respect to the general introduction of heteroatoms at a sterically hindered $C(sp^3)$ center.

4.2 Results and Discussion

To investigate the nature of boron-ate complexes, the initial electrochemical assessments of hindered trifluoroborate substrates **a** and **b** were made using cyclic voltammetry under an inert atmosphere (Scheme 4.1A). This showed two distinct peaks representing oxidation events. Additional analysis with differential pulse voltammetry and square wave voltammetry supported two distinct oxidation events with improved resolution (see the Sect. 4.4 for details) [13]. Thus, the collected evidence clearly demonstrated that two successive oxidation events of organoboron compounds occur under electrochemical conditions. A more quantitative evaluation of the event was performed using a thin-layer electroanalysis microchip (TEAM) (Scheme 4.1B) [14]. A model substrate, trifluoroborate **c**, underwent the first oxidation event with the peak potential of 1.1 V. A quantitative measurement of the number of electrons transferred per molecule, *n*, which, in this case, equals 1, indicated SET. On the other hand, the electrolysis at the second oxidation peak of **c** at 1.5 V gave the n value converging to 2, indicating two distinct SET events under the given potential.

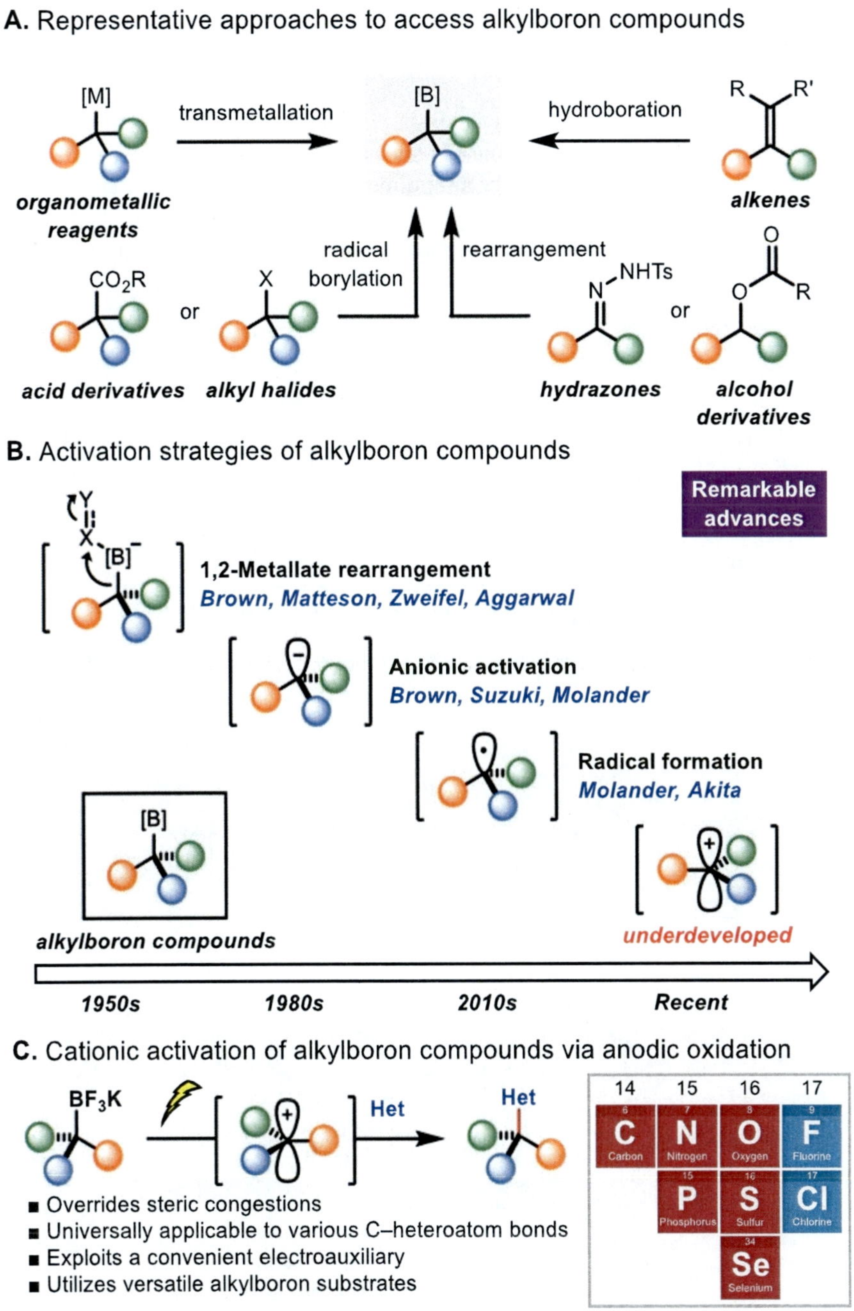

Fig. 4.1 Preparation and activation of alkylboron compounds

Based on these electrochemical analyses, we hypothesized that the two successive SET would induce the postulated carbocation intermediate necessary for the desired bond formation.

Based on the electrochemical analysis, a strategy for C(sp^3)–B bond activation was established, which ultimately allowed the introduction of a wide range of heteroatoms (Scheme 4.1C). The multidimensional reaction parameter evaluation was summarized as following results: (a) external additives, such as oxidants, bases, or dehydrating agents, were demonstrated to be nonessential; (b) an inexpensive

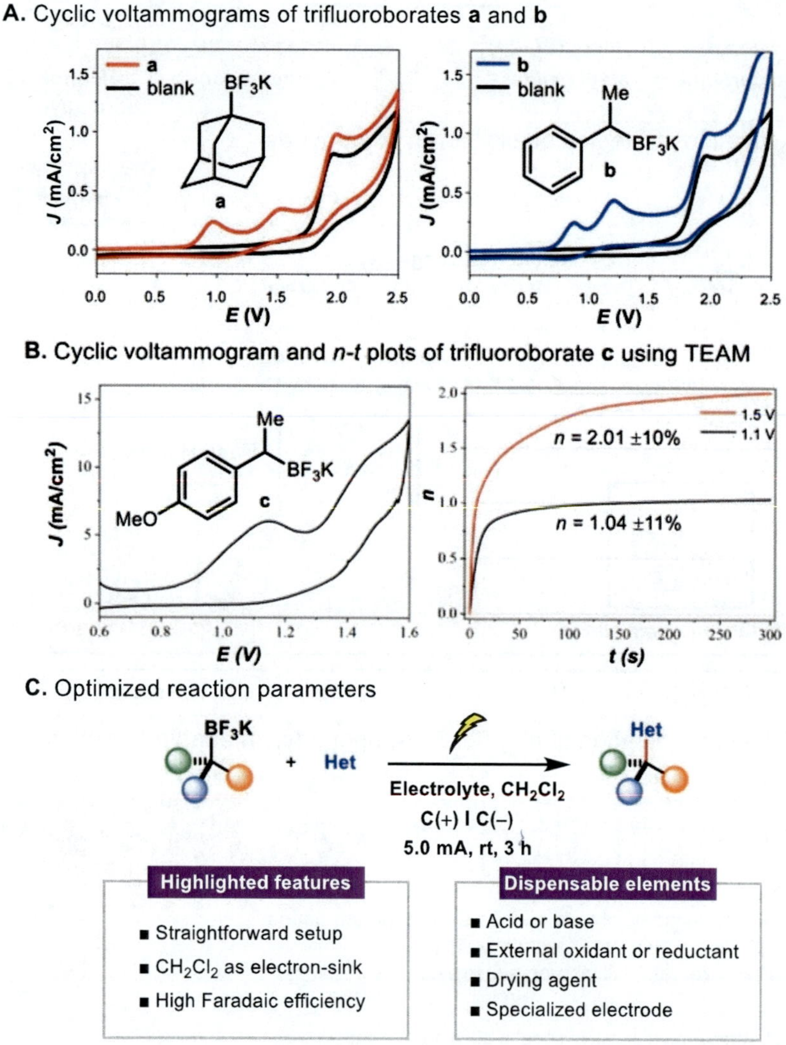

Scheme 4.1 Identification of reaction conditions on the basis of electrochemical analysis

graphite electrode gave the desired coupling product with the greatest efficiency; (c) dichloromethane as the reaction solvent, which also acts as an electron sink [4d, 15], elicited the best performance (see the Sect. 4.4 for details); and (d) the best yield could be obtained within 3 h for a 0.2 mmol scale reaction setup. From an electrochemical point of view, the faradaic efficiency of the system was generally in the 40–80% range. In addition, a current of 5.0 mA, which resulted in a 5.0–6.0-V initial cell voltage between the anode and cathode, was optimal, although deviation from 5.0 mA did not significantly influence the reaction yield. In summary, the $C(sp^3)$–heteroatom bond formation under constant current in an undivided cell was generally achieved with a straightforward reaction setup.

After evaluation of reaction parameters, the strategy was applied to the $C(sp^3)$–O bond formation with an alcohol or a carboxylic acid (Table 4.1). In the presence of an alcohol nucleophile, sterically hindered benzyl trifluoroborate substrates smoothly underwent ether formation (Table 4.1A, **1–15**). Secondary benzylic substrates presented the corresponding ethers with cyclohexanol regardless of the steric effects around the reaction center (**1–3**). In different circumstances, variations in the electronic nature of the substrates significantly affected the reaction yields (**4–7**). As the aromatic ring became more electron-rich, the desired product formation increased distinctly. The tendency demonstrates the involvement of a reactive intermediate with carbocationic property. Other benzyl compounds containing multiple arene rings (**8**, **10**), an extended π-system (**11**), or a tertiary reaction center (**12**) also underwent facile alkoxylation under the given conditions. In the case of the alcohol nucleophiles, oxidatively labile 1,4-benzenedimethanol (**13**), and sterically demanding secondary (**14**) or tertiary (**15**) carbinol furnished the coupling products with high efficiency.

Secondary and tertiary nonbenzyl trifluoroborate salts were also viable substrates for the desired transformation (Table 4.1A, **16–28**). Sterically hindered α-tertiary or β-quaternary dialkyl ether could be obtained in a straightforward manner using a variety of alcohol substrates. Substrates based on norbonyl or adamantyl backbone underwent etherification with high efficiency (**16–19**, **22**, **28**). Gratifyingly, alcohols containing sensitive functional groups such as an alkyne, an acetal, or a chloride performed the desired transformation (**17–19**). Other tertiary alkyl trifluoroborates with cyclic or linear aliphatic chains also underwent the desired transformation (**20**, **21**, **23–27**). Surprisingly, the generation of a bridgehead carbocation was identified, providing the desired product (**23**). Last, extremely sterically hindered quaternary centers could be connected through an oxygen linker (**28**).

Next, the reactivity scope of weaker oxygen nucleophiles, carboxylic acids, was evaluated to afford hindered esters (Table 4.1B). On the nucleophile side, a variety of aliphatic and aromatic carboxylic acids could be directly introduced to the $C(sp^3)$ center of the organoboron reagents to furnish a variety of alkyl esters (**29–44**). Notably, carboxylic acids vulnerable to oxidative conditions could be utilized as reaction partners without competitive generation of reactive species from the acids (**29–35**, **43**) [16]. Furthermore, reactive α,β-unsaturated carbonyl systems (**38**, **39**) and biomedically important heterocycles (**40–44**) were tolerated. Remarkably, in the

Table 4.1 Synthesis of sterically congested ethers and esters by the introduction of oxygen nucleophiles

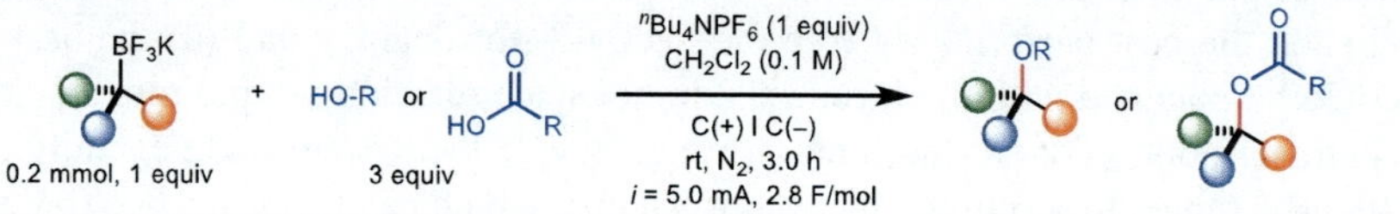

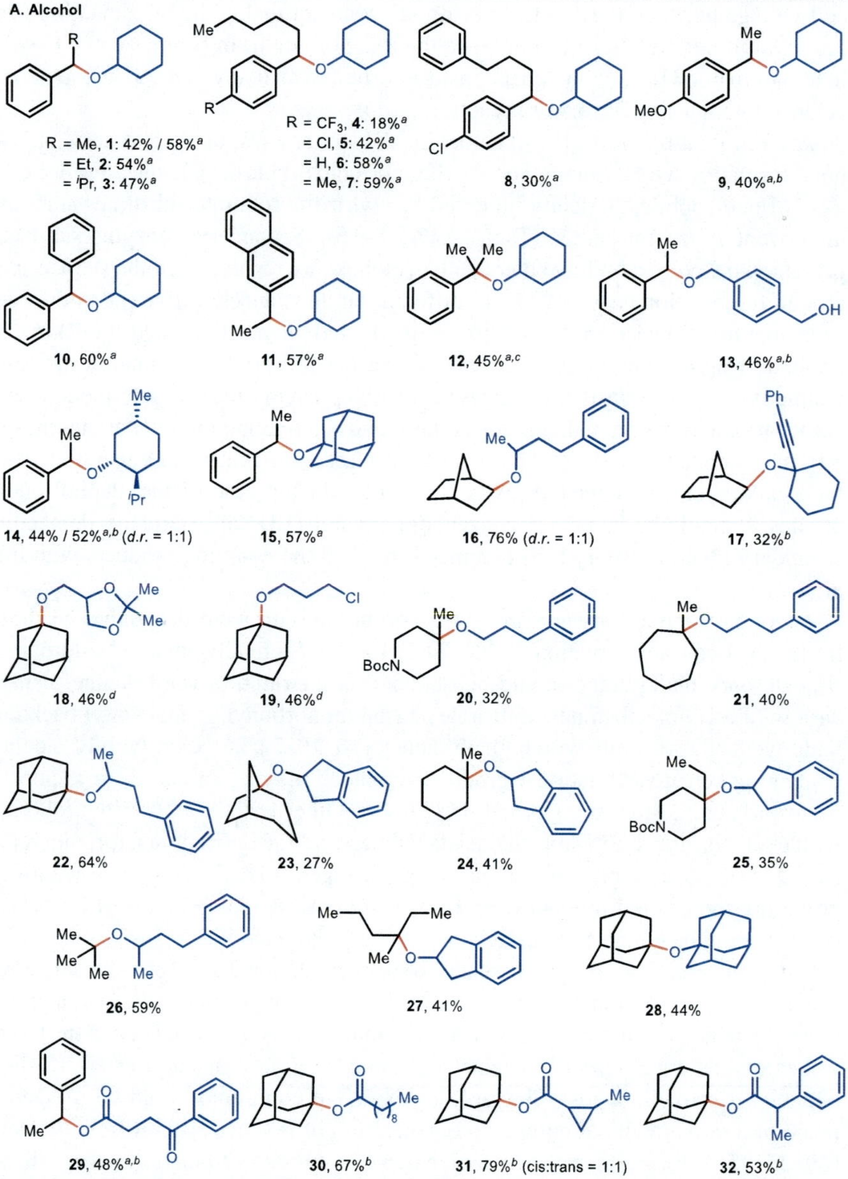

(continued)

Table 4.1 (continued)

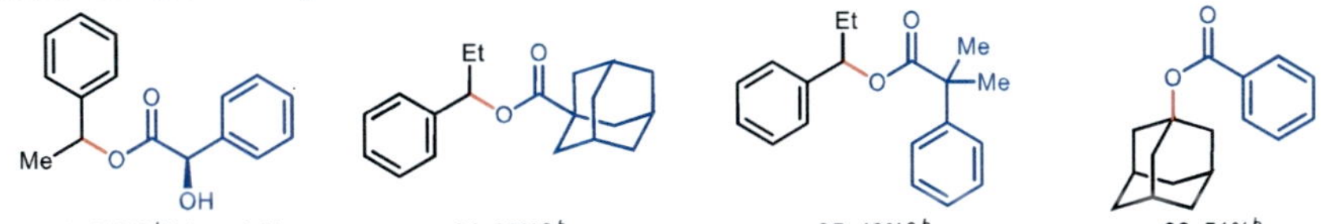

33, 66%[a,b] (*d.r.* = 1:1) **34**, 38%[a,b] **35**, 43%[a,b] **36**, 51%[b]

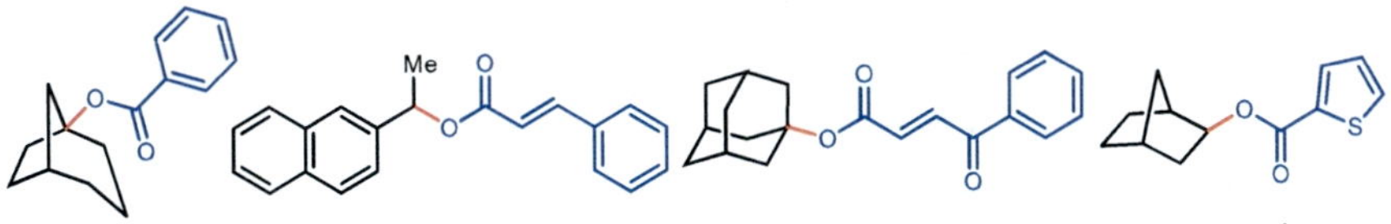

37, 33%[b] **38**, 56%[a, b] **39**, 81%[b] **40**, 67%[b]

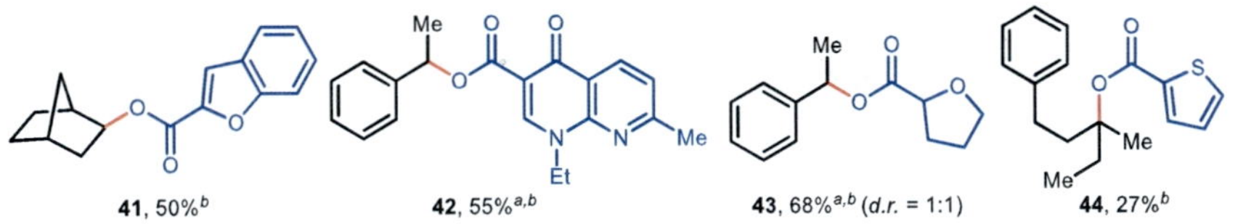

41, 50%[b] **42**, 55%[a,b] **43**, 68%[a,b] (*d.r.* = 1:1) **44**, 27%[b]

[a] $^{n}Bu_4NClO_4$ (1 equiv) was used as an electrolyte instead of $^{n}Bu_4NPF_6$ (1 equiv)
[b] The reaction was run for 2.0 h instead of 3.0 h
[c] The reaction was run for 1.5 h instead of 3 h
[d] 6 equivalent of the nucleophile was used
$^{n}Bu_4NClO_4$, tetra-*n*-butylammonium perchlorate; rt, room temperature; *d.r.*, diastereomeric ratio; *i* represents constant current

case of mandelic acid containing a hydroxyl group, the carbocation generated from alkyl boron reagents selectively reacted with the carboxylic acid moiety (**33**).

The reactivity of alkyl trifluoroborate salts was explored to build other types of sterically crowded molecules (Table 4.2). Emphatically, the generated reactive intermediate could form a covalent bond with virtually all nonmetallic elements except for inert elements. A variety of electron-rich aromatic compounds enabled C(sp^3)–C(sp^2) bond formation (Table 4.2A, **45–47**). Interestingly, bioactive indole derivatives with decreased electron densities could react with sterically hindered C(sp^3) building blocks (**48**, **49**). In addition, the regioselective formation of C3-alkylated indoles supported that a carbocation intermediate is responsible for bond formation rather than a radical species [17]. Several nitrogen nucleophiles were also viable reaction counterparts for this strategy (Table 4.2B, **50–61**). Sterically hindered C(sp^3)–N bonds of sulfonamides and a phosphoramidate could be achieved efficiently (**50–54**). Pharmaceutically potent azole derivatives could participate in C(sp^3)–N bond formation in good yields (**55–57**) [18]. Importantly, alkylation of carbamates was also feasible, which presented highly substituted carbamate products (**58–60**). In principle, access to complex amines is possible through deprotection of the *tert*-butyloxycarbonyl protecting group from the carbamate products (**58**, **59**). Furthermore, a redox-sensitive acylhydrazide was selectively alkylated on the terminal nitrogen, showing the mildness of the protocol (**61**).

Table 4.2 Introduction of other heteroatoms into an sp^3-hybridized carbon atom

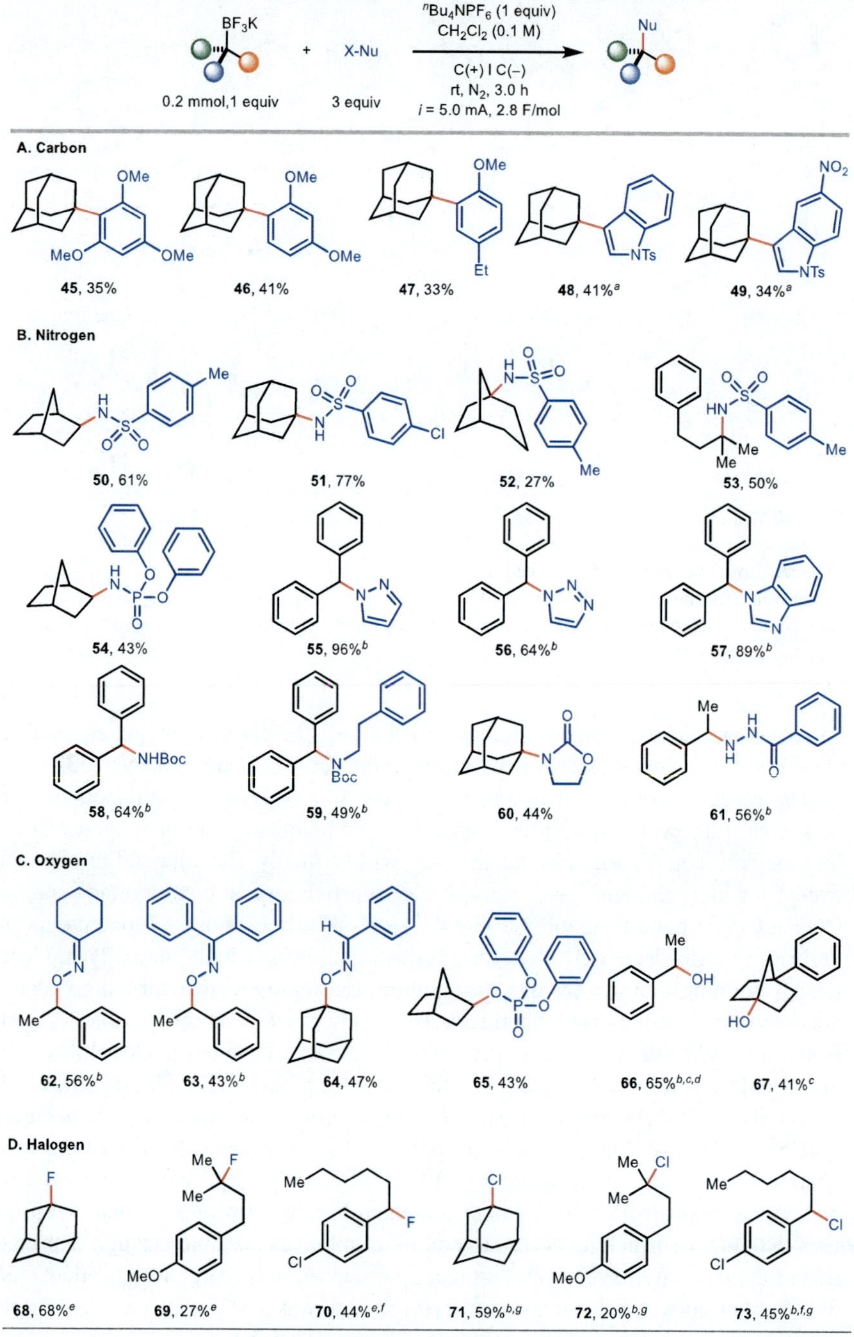

(continued)

Table 4.2 (continued)

E. Other pnictogen & chalcogen

74, 47%[b,h]

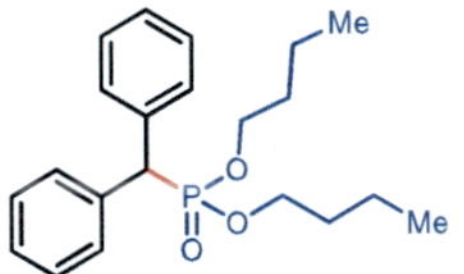

75, 52%[b,i]

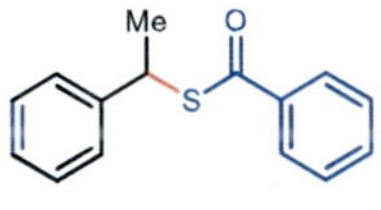

76, 53%[b]

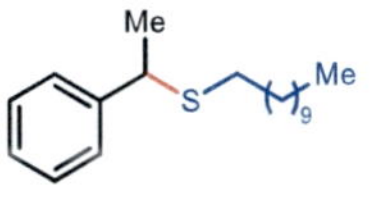

77, 43%[b,d]

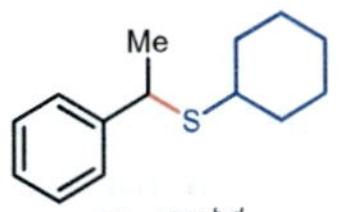

78, 47%[b,d]

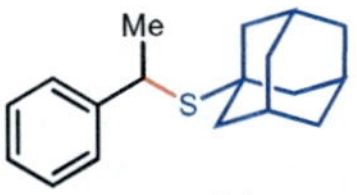

79, 40%[b,d]

80, 45%[b,d,j]

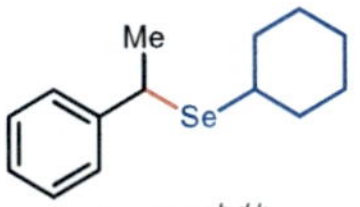

81, 41%[b,d,k]

[a]The reaction was set up with alkyl trifluoroborate (0.1 mmol), nucleophile (2 equiv), and $^{n}Bu_4NPF_6$ (2 equiv) in CH_2Cl_2 (0.05 M) with 3.0 mA of electric current
[b]$^{n}Bu_4NClO_4$ (1 equiv) was used as an electrolyte instead of $^{n}Bu_4NPF_6$ (1 equiv)
[c]H_2O (0.1 mL) was used as a nucleophile
[d]The reaction was run for 1.5 h instead of 3.0 h
[e]KHF_2 (2 equiv) was used as a nucleophile with 18-crown-6 (2 equiv)
[f]The reaction was run for 2.0 h instead of 3.0 h
[g]KCl (2 equiv) was used as a nucleophile with 18-crown-6 (2 equiv)
[h]Triethyl phosphite (3 equiv) was used as a source of heteroatom
[i]Tributhyl phosphite (3 equiv) was used as a source of heteroatom
[j]1,2-Dimethyldiselane (3 equiv) was used as a nucleophile
[k]1,2-Dicyclohexyldiselane (3 equiv) was used as a nucleophile
Ts, tosyl; Boc, *tert*-butyloxycarbonyl

This strategy has also been applied to other types of oxygen nucleophiles that are not alcohols or carboxylic acids (Table 4.2C). Medicinally potent oxime derivatives could be synthesized by the introduction of alkyl groups on the oxygen (**62–64**). Aliphatic/aromatic ketoximes and aldoximes were competent in the formation of $C(sp^3)$–O bonds, enabling the preparation of related alkoxyamines with academic and industrial significance via post-functionalization. Furthermore, a mixed phosphate triester could be synthesized by the reaction of a weak oxygen nucleophile (**65**). The transformation of a hindered borate to the corresponding alcohol without using any chemical oxidants occurred (**66, 67**).

Even halogenated products could be obtained by using potassium hydrogen difluorides (KHF_2) or potassium chloride (KCl) as halogen sources (Table 4.2D, **68–73**). In addition, the established methodology realized the bond formation reactions with pnictogen and chalcogen elements in the third and fourth rows (Table 4.2E). Trialkyl phosphites participated in a dealkylative $C(sp^3)$–P bond formation to give hindered phosphonate products (**74, 75**). Sulfur-containing compounds, such as thioacid or thiols, also furnished a variety of organosulfur compounds (**76–79**). Finally, an exploration of this method could extend to the synthesis of selenides (**80, 81**). Of note, the sources of group 16 nucleophiles are in their native form, which is inherently the most stable and therefore the easiest to manipulate. Moreover, the oxidatively labile products, thioethers and selenides, could be obtained in moderate yields by adjusting the reaction time. Overall, the introduction of heteroatoms to the carbon framework via a unified synthetic platform was realized to afford new heteroatom-containing scaffolds.

The robust and practical nature of this synthetic protocol was further demonstrated by the preparation of complex molecules based on a steroidal backbone structure (Scheme 4.2A). The hydroxyl groups of complex steroid natural products or their derivatives have successfully participated in these bond formations (**82–85**). Moreover, a complex alkyl trifluoroborate originating from enoxolone could undergo a chemoselective transformation to afford valuable products with a heteroatom substituent, which enables the late-stage $C(sp^3)$–B functionalization (**86**, **87**). Additionally, the practicality of the method was evaluated based on a preparatory-scale reaction setup (Scheme 4.2B). When the reaction scale was increased tenfold, the reaction yield remained virtually unchanged, although a longer reaction time was essential due to the supply of electrons required per molecule. Finally, to realize a more practical operation of the strategy, a more easily accessible organoboron substrate was directly tested as a reaction counterpart (Scheme 4.2C). The addition of fluoride sources into the reaction system was anticipated to generate in situ generation of alkyl fluoroboronate derivatives, which could avoid precipitation of boronate salts. As a result, the step-economical pathway starting from alkyl pinacol boronic esters, the immediate precursors of trifluoroborate, gave the desired product with comparable efficiency. Overall, the power of the protocol as a synthetic tool has been demonstrated in multiple ways.

Next, a number of experimental observations were collected to elucidate the detailed reaction pathways. First, the carbocationic nature of reactive intermediates could be clearly confirmed (Scheme 4.3). For benzyl trifluoroborate substrates, it was observed that a dependence of the reaction efficiency on the para substituent effects of the arene (Scheme 4.3A). While benzyl trifluoroborates containing electron-rich arene underwent efficient C–O bond formation, electron-deficient benzyl trifluoroborates suffered from competing side reactions, mainly protodeborylation. Besides, a migratory aptitude of the reaction center in alkylboron compounds indicated the participation of a carbocation intermediate, a species that commonly rearranges to form a more stable isomer (Scheme 4.3B). When bridgehead carbocations occurred, the birth of more stable cations with the ring-opening of strained carbocycles was detected to give rearrangement products (Scheme 4.3B, (**1**), (**2**)) [4d, 20]. Also, a cation rearrangement from a tertiary position to α-heteroatom-stabilized position via a 1,2-hydride shift was observed (Scheme 4.3B, (**3**)) [4d, 21].

Additional electrochemical experiments presented mechanistic information regarding the activation of $C(sp^3)$–B bonds (Scheme 4.4). First, electrolysis experiments in a divided cell described that product formation took place exclusively in the anodic chamber (Scheme 4.4A). In a divided cell, both the trifluoroborate **a** and the alcohol in the anodic chamber were required for the successful formation of the product (Scheme 4.4A, entries 1 and 2). The results indicate that the anodic oxidation is responsible for the observed reactivity, $C(sp^3)$–heteroatom bond formation. Notably, the removal of the membrane from an identical divided cell

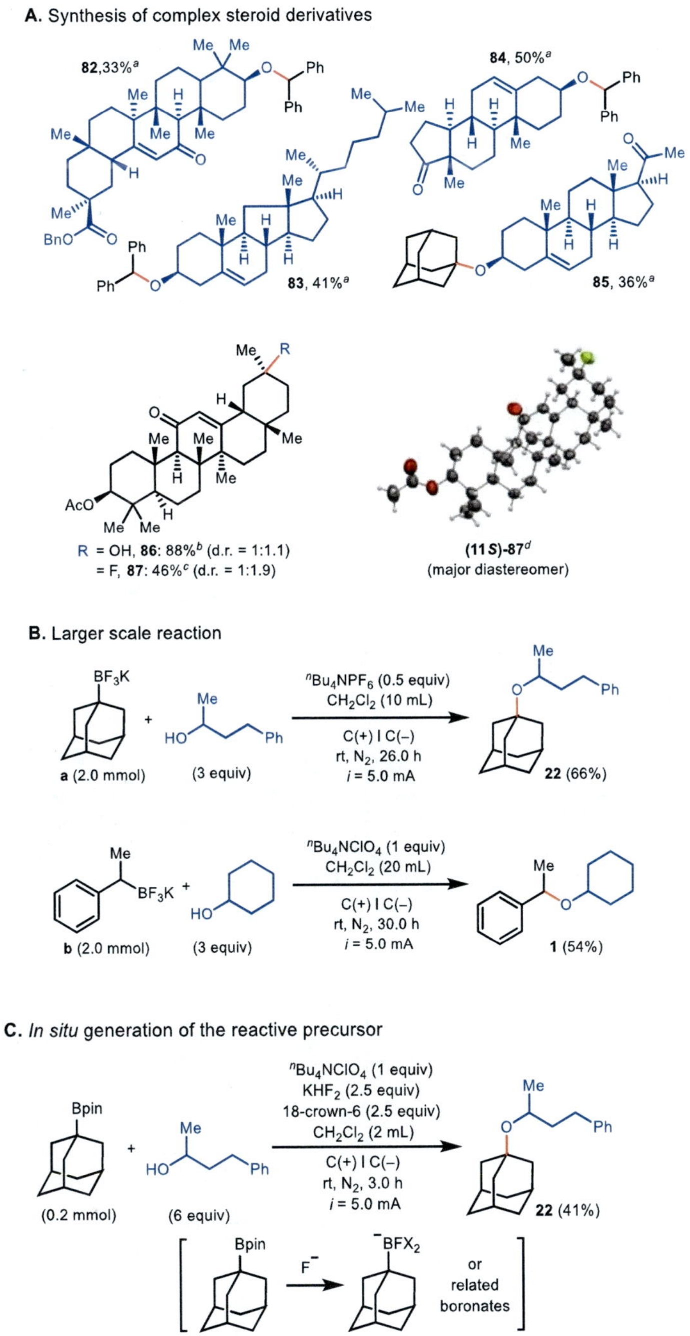

Scheme 4.2 Synthetic applications of the developed strategy

A. Substituent effects in substituted aromatic rings

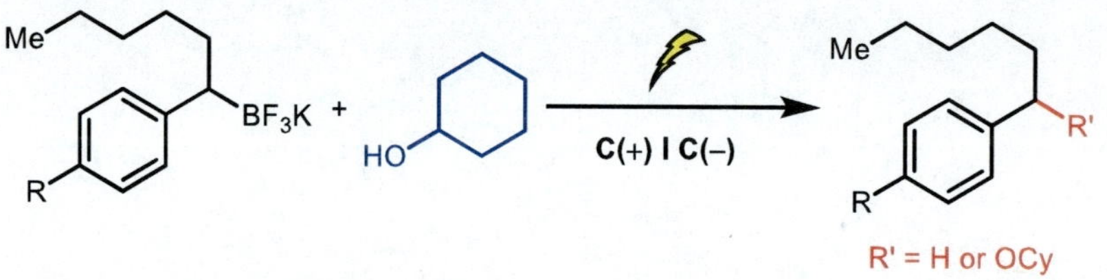

entry	substituent constant (σ_{para})	yield (R' = H) (%)	yield (R' = OCy) (%)
1 (R = CF_3)	+0.54	40	18
2 (R = Cl)	+0.23	22	42
3 (R = H)	+0.00	13	58
4 (R = Me)	–0.17	<1	59

B. Carbocation rearrangement via anodic oxidation of C(sp^3)–B bond[a]

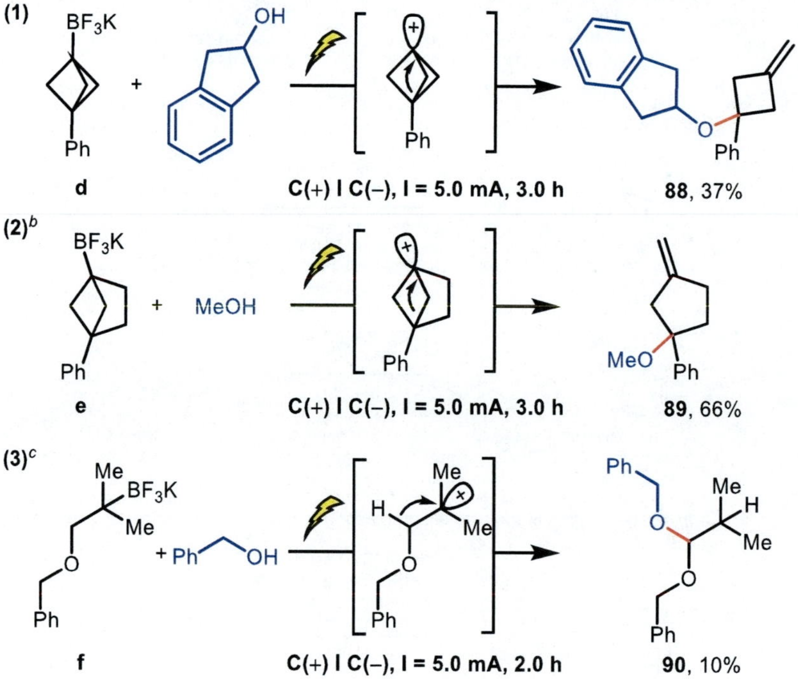

Scheme 4.3 Evidences for the generation of carbocation intermediates

assembly led to an increased reaction yield (Scheme 4.4A, entry 3). It is speculated that the high reaction potential caused by membrane resistance facilitated the decomposition of both the reactant and the product in the divided electrochemical cell. In conjunction with other experiments, a detailed mechanistic scenario could be drawn from a constant potential electrolysis experiment using the reaction of trifluoroborate **a** and 4-phenyl-2-butanol (Scheme 4.4B). With the amount of total charge transfer remaining unchanged, the reaction yields were measured by varying the applied potential. When the applied potential increased to 1.5 V (vs Ag/AgCl in 3.0 M NaCl), a second anodic peak potential of the trifluoroborate **a** (Scheme 4.1A), an abrupt increase in the product yield, was observed. Simultaneously, a decrease in the yields of the side products, such as the protodeborylation and fluorination products, was obtained. Therefore, for the effective generation of desired products, the cell potential should be required to reach the second anodic peak potential, at which point the desired pathway prevails. Finally, the kinetic investigation of alkyl ether formation exhibited a positive rate order with respect to the applied current (Scheme 4.4C). On the contrary, the concentration of alkyl trifluoroborates or alcohols did not significantly affect the reaction rate. The reaction rate profiles of these systems imply that a heterogeneous electron-transfer between an electrode and alkyl trifluoroborates could be the rate-determining step (see the Sect. 4.4 for details).

Based on the evidences collected from multiple mechanism experiments, a plausible mechanism was proposed (Scheme 4.5). First, alkyl trifluoroborates undergo two sequential single-electron transfers with a concomitant cleavage of the C–B bond via an electron transfer–chemical reaction–electron transfer (ECE) mechanism at the anode [9e, 12]. One-electron oxidation of an alkyl trifluoroborate provides a transient radical species, which can be rapidly followed by BF_3 removal and additional single-electron transfer to generate a carbocation. The electrogenerated carbocation should proceed with C(sp^3)–heteroatom bond formation to furnish the product. On the cathode, a reduction process of dichloromethane should simultaneously balance the charge state of the overall system together with a minor pathway originating from the reduction of nucleophiles [4e, 15]. However, the proton reduction of nucleophiles can be varied depending on their nature (see the Sect. 4.4 for details).

4.3 Conclusion

In summary, organoboron compounds have drawn attention from a synthetic perspective to enable the construction of a wide range of C(sp^3)–heteroatom bonds, providing an unprecedented synthetic utility. The desired reactivity is derived from two consecutive electrochemically driven single-electron oxidations of organoboron compounds, which affords carbocations, the intermediates with elevated reactivity. As a result, our strategy has realized a general synthetic platform that can react electrogenerated carbocation with virtually all nonmetallic elements. This method is suitable for the formation of C(sp^3)–heteroatom bonds at a sterically demanding reaction center and is applicable to the synthesis of complex molecules.

A. Divided cell experiments

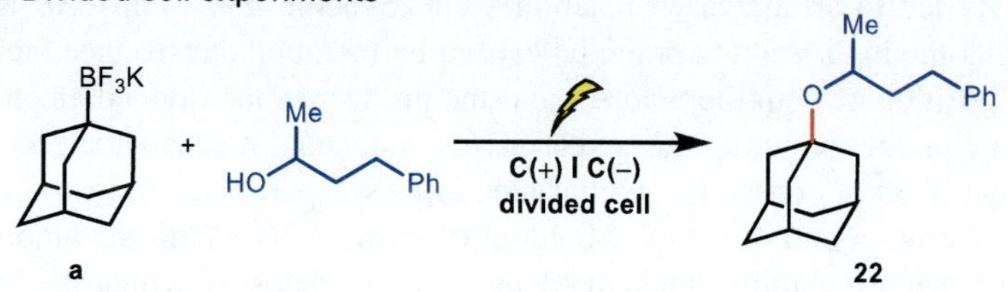

	anodic chamber	cathodic chamber	membrane	current (mA)	time (h)	yield (%) anodic chamber	yield (%) cathodic chamber
1[a]	**a**, ROH	ROH	O	3	20	3 (8)[d]	-
2[b]	ROH	**a**, ROH	O	3	20	-	-
3[c]	**a**, ROH	ROH	–	3	20	13	-

B. Constant potential electrolysis of trifluoroborate **a**

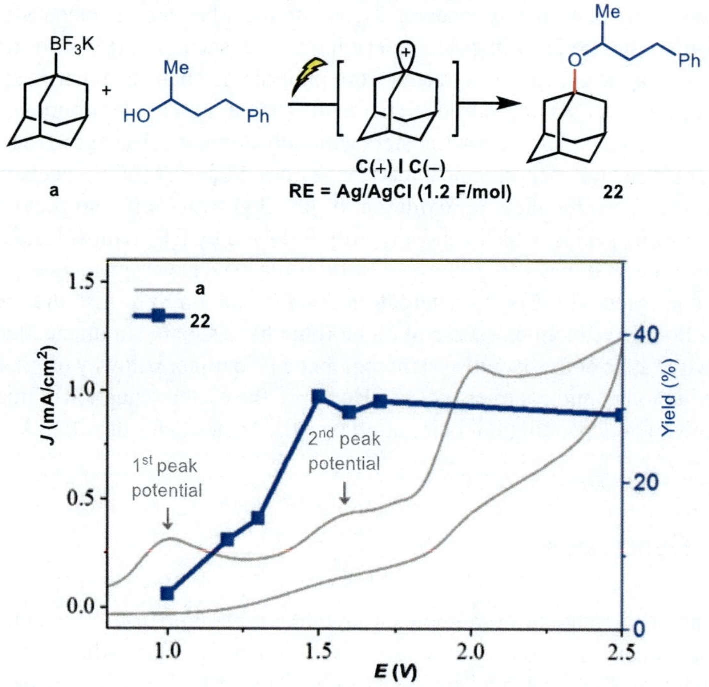

Scheme 4.4 Further electrochemical investigations of the reaction mechanism

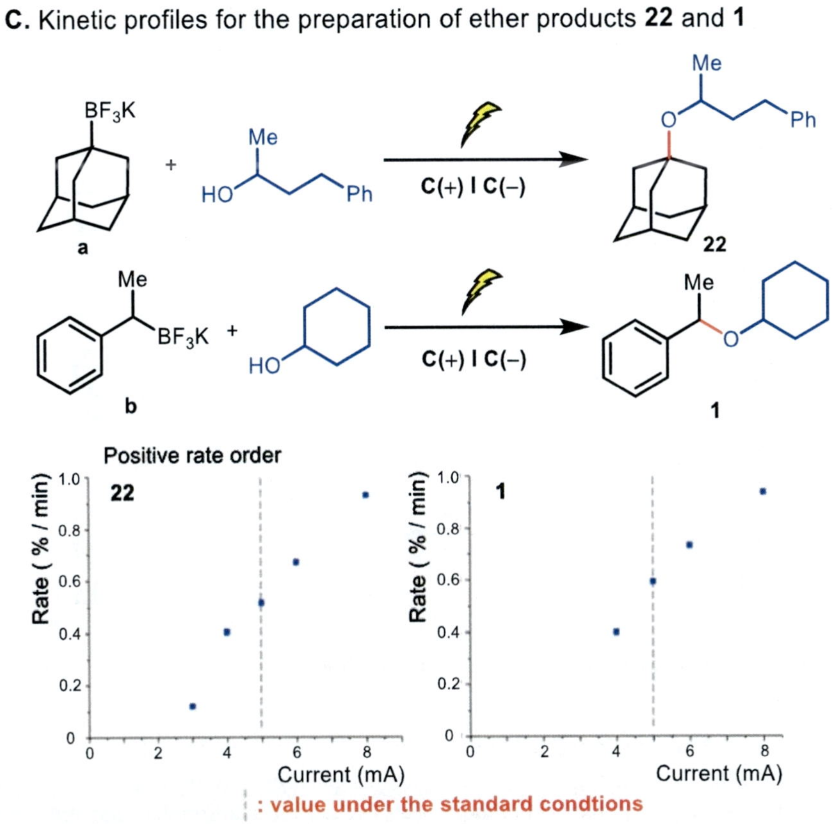

Scheme 4.4 (continued)

4.4 Experimental Section

4.4.1 General Information

Dichloromethane (CH_2Cl_2), 1,4-dioxane, and toluene were dried using a Pure Solv solvent purification system. Anhydrous *N*,*N*-dimethylformamide (DMF), acetonitrile (CH_3CN), and dimethylsulfoxide (DMSO) were purchased from Sigma-Aldrich. All chemicals were purchased from commercial sources (Sigma-Aldrich, Alfa Aesar, TCI, or Strem) and used without further purification. Deuterated compounds were purchased from Cambridge Isotope Laboratories, Inc. and Sigma-Aldrich Corporation. Nuclear magnetic resonance (NMR) spectra were recorded in $CDCl_3$, DMSO-d_6, and acetone-d_6 on a Varian 400 NMR (400 MHz), 500 NMR (500 MHz), Bruker 500 NMR (500 MHz), and Bruker 600 NMR (600 MHz) spectrometers. ^1H NMR and ^{13}C NMR chemical shifts were referenced to the residual solvent signal ($CHCl_3$ in $CDCl_3$: δ 7.26 ppm for ^1H, δ 77.16 ppm for ^{13}C; $(CH_3)_2CO$ in $(CD_3)_2CO$: δ 2.05 ppm

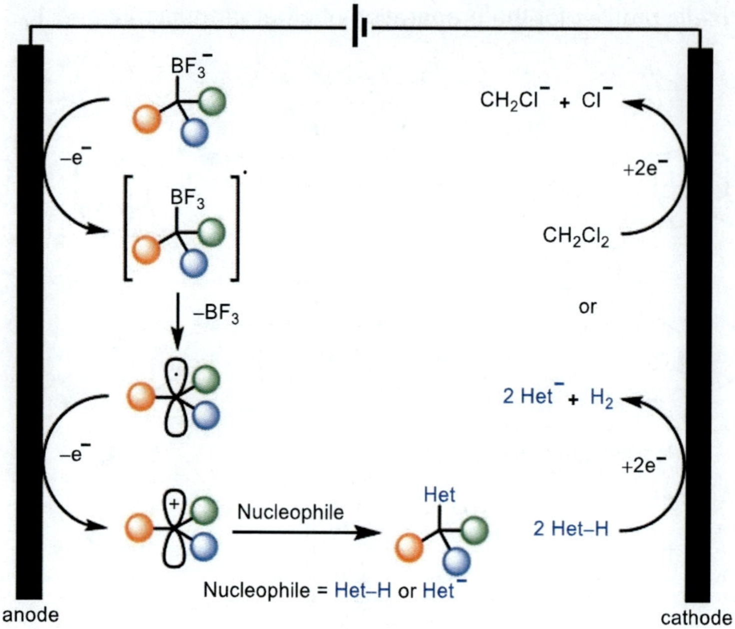

Scheme 4.5 Proposed mechanism for the overall electrochemical process

for ^{1}H, δ 29.84 ppm and δ 206.26 ppm for ^{13}C; $(CH_3)_2SO$ in $(CD_3)_2SO$: δ 2.50 ppm for ^{1}H, δ 39.52 ppm for ^{13}C). The ^{13}C signal of the carbon atoms directly bonded to a boron atom was not observed in most cases because of quadrupolar relaxation. ^{19}F NMR chemical shifts were referenced to external trifluorotoluene (δ −3.72 ppm). ^{11}B NMR chemical shifts were referenced to external $BF_3{\cdot}OEt_2$ (0.0 ppm). ^{31}P NMR chemical shifts were referenced to an external triphenylphosphine (δ −6.0 ppm). Chemical shifts are reported in ppm and coupling constants are given in Hz. The following abbreviations were used to explain multiplicities: s = singlet, d = doublet, t = triplet, q = quartet, p = pentet, m = multiplet, br = broad. Gas chromatography (GC) was carried out using a GC-2030 (Shimadzu) equipped with an Rxi®-5Sil MS column and a flame ionization detector (FID). High-resolution mass spectrometry (HRMS) was performed at the Organic Chemistry Research Center at Sogang University using the ESI method, Department of Chemistry at Seoul National University using the ESI method, and the Daegu Center of Korea Basic Science Institute using the EI method. The crystal structure was determined by single-crystal X-ray diffractometer at the Western Seoul Center of Korea Basic Science Institute. All electrochemical measurements were performed with CHI 660 and 750 potentiostat (CH Instruments, TX, USA) and dual display potentiostat (DJS-292B and DJS-292C, China). Unless otherwise noted, all reactions were performed under nitrogen. Reactions were monitored by thin-layer chromatography (TLC) on EMD Silica Gel 60 F254 plates, and visualized either using UV light (254 nm) or by staining with potassium permanganate ($KMnO_4$) and heating.

4.4.2 Substrate Preparation

Spectral Data Matched that Reported in the Literature: 4,4,5,5-Tetramethyl-2-(1-phenylethyl)-1,3,2-dioxaborolane (**S1a**) [22], 4,4,5,5-tetramethyl-2-(1-phenylpropyl)-1,3,2-dioxaborolane (**S2a**) [22], 4,4,5,5-tetramethyl-2-(2-methyl-1-phenylpropyl)-1,3,2-dioxaborolane (**S3a**) [22], 2-benzhydryl-4,4,5,5-tetramethyl-1,3,2-dioxaborolane (**S4a**) [23], 4,4,5,5-tetramethyl-2-(1-(naphthalen-2-yl)ethyl)-1,3,2-dioxaborolane (**S5a**) [23], 2-(1-(4-methoxyphenyl)ethyl)-4,4,5,5-tetramethyl-1,3,2-dioxaborolane (**S6a**) [23], 4,4,5,5-tetramethyl-2-(1-methylcyclohexyl)-1,3,2-dioxaborolane (**S7a**) [24], (3*S*,4a*R*,6a*R*,6b*S*,8a*R*,11*S*,12a*R*,14a*R*,14b*S*)-4,4,6a,6b,8a,11,14b-heptamethyl-14-oxo-11-(4,4,5,5-tetramethyl-1,3,2-dioxaborolan-2-yl)-1,2,3,4,4a,5,6,6a,6b,7,8,8a,9,10,11,12,12a,14,14a,14b-icosahydropicen-3-yl acetate (**S8a**) [24], 4,4,5,5-tetramethyl-2-(2-phenylpropan-2-yl)-1,3,2-dioxaborolane (**S9a**) [24], 2-(tert-butyl)-4,4,5,5-tetramethyl-1,3,2-dioxaborolane (**S10a**) [25], 2-(bicyclo[3.2.1]octan-1-yl)-4,4,5,5-tetramethyl-1,3,2-dioxaborolane (**S11a**) [25], 4,4,5,5-tetramethyl-2-(3-phenylbicyclo[1.1.1]pentan-1-yl)-1,3,2-dioxaborolane (**S12a**) [26], 4,4,5,5-tetramethyl-2-(4-phenylbicyclo[2.1.1]hexan-1-yl)-1,3,2-dioxaborolane (**S13a**) [26], 4,4,5,5-tetramethyl-2-(4-phenylbicyclo[2.2.1]heptan-1-yl)-1,3,2-dioxaborolane (**S14a**) [26], 2-(bicyclo[2.2.1]heptan-2-yl)-4,4,5,5-tetramethyl-1,3,2-dioxaborolane (**S15a**) [27], 2-(adamantan-1-yl)-4,4,5,5-tetramethyl-1,3,2-dioxaborolane (**S16a**) [27], 2-(4-(4-methoxyphenyl)-2-methylbutan-2-yl)-4,4,5,5-tetramethyl-1,3,2-dioxaborolane (**S23a**) (Fig. 4.2) [28].

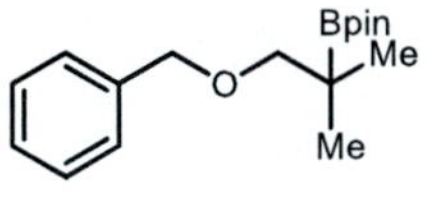

2-(1-(Benzyloxy)-2-methylpropan-2-yl)-4,4,5,5-tetramethyl-1,3,2-dioxaborolane (S17a)

Alkylboronic pinacol ester **S17a** was synthesized via a previously reported procedure [27]. Purification by flash column chromatography (silica gel, hexanes:CH_2Cl_2 = 7:3) afforded **S17a** (0.52 g, 44%) as a colorless oil; $R_f = 0.2$ (hexanes:CH_2Cl_2 = 4:1); ^{1}H NMR (500 MHz, $CDCl_3$): δ 7.36–7.21 (m, 5H), 4.51 (s, 2H), 3.31 (s, 2H), 1.22 (s, 12H), 0.98 (s, 6H); ^{13}C NMR (126 MHz, $CDCl_3$): δ 139.4, 128.3, 127.3, 127.2, 83.2, 79.7, 73.2, 24.8, 22.0; ^{11}B NMR (193 MHz, $CDCl_3$): δ 34.3; HRMS (ESI) *m*/*z*: $[M+Na]^+$ calcd for $C_{17}H_{27}BO_3Na$ 313.1946; found 313.1941.

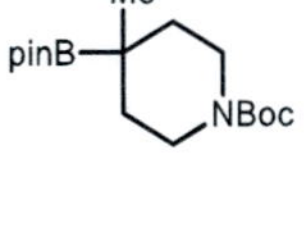

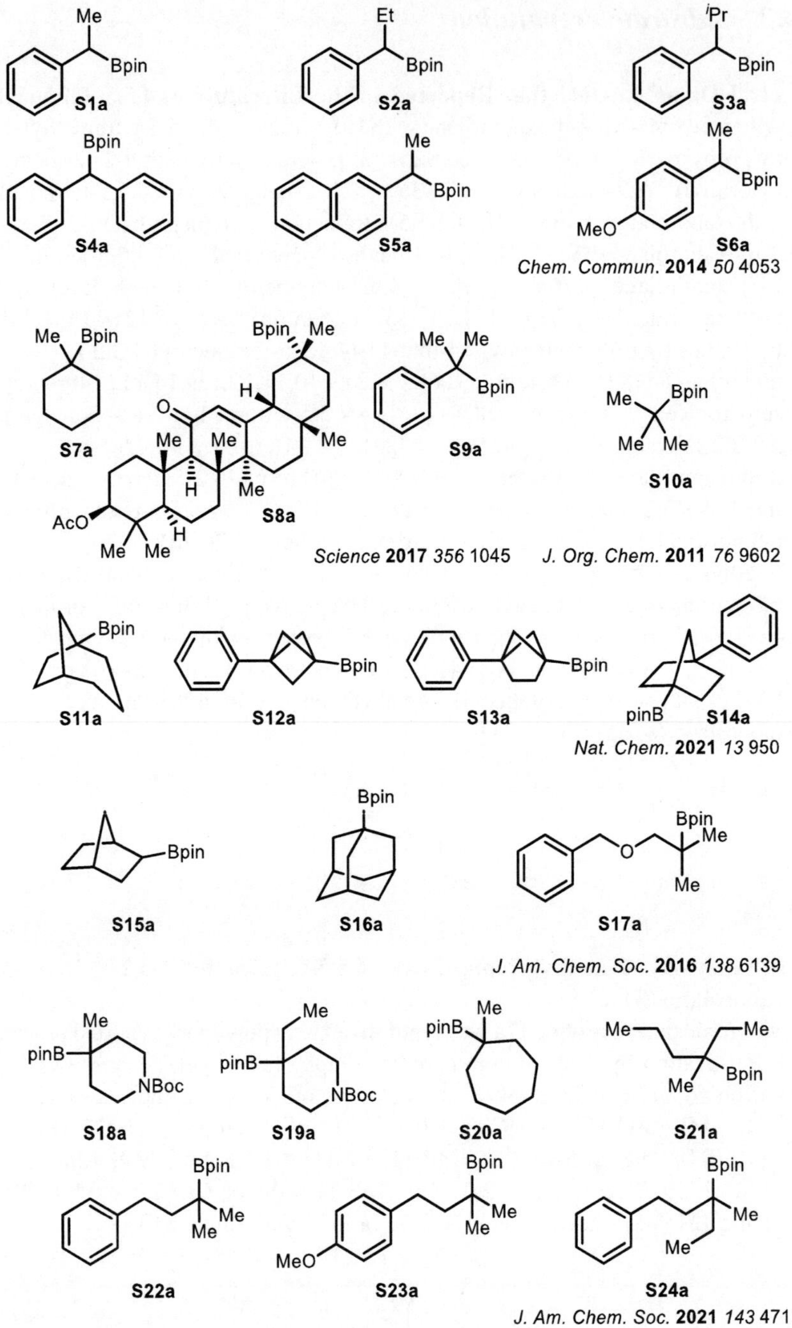

Fig. 4.2 List of alkylboronic pinacol esters for their preparation

***tert*-butyl 4-methyl-4-(4,4,5,5-tetramethyl-1,3,2-dioxaborolan-2-yl)piperidine-1-carboxylate (S18a)**

Alkylboronic pinacol ester **S18a** was synthesized via a previously reported procedure [28]. Purification by flash column chromatography (silica gel, hexanes:EtOAc = 20:1) afforded **S18a** (0.99 g, 61%) as a white solid; m.p. 64–66 °C; $R_f = 0.3$ (hexanes:EtOAc = 10:1); ^{1}H NMR (400 MHz, $CDCl_3$) δ 3.94 (m, 2H), 2.82–2.63 (m, 2H), 1.78 (m, 2H), 1.44 (s, 9H), 1.22 (s, 12H), 1.18–1.05 (m, 2H), 0.93 (s, 3H); ^{13}C NMR (101 MHz, $CDCl_3$) δ 155.2, 83.3, 79.2, 43.3, 35.9, 28.6, 25.1, 24.8; ^{11}B NMR (161 MHz, $CDCl_3$): δ 34.6; HRMS (ESI) *m/z*: $[M+H]^+$ calcd for $C_{17}H_{32}BNO_4H$ 326.2503; found 326.2490.

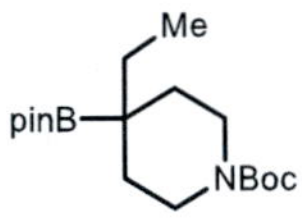

***tert*-butyl 4-ethyl-4-(4,4,5,5-tetramethyl-1,3,2-dioxaborolan-2-yl)piperidine-1-carboxylate (S19a)**

Alkylboronic pinacol ester **S19a** was synthesized via a previously reported procedure [28]. Purification by flash column chromatography (silica gel, hexanes:EtOAc = 30:1) afforded **S19a** (0.95 g, 56%) as a white solid; m.p. 64–66 °C; $R_f = 0.3$ (hexanes:EtOAc = 10:1); ^{1}H NMR (400 MHz, $CDCl_3$) δ 3.97 (m, 2H), 2.80–2.71 (m, 2H), 1.88–1.79 (m, 2H), 1.44 (s, 9H), 1.32 (q, $J = 7.5$ Hz, 2H), 1.24 (s, 12H), 1.05 (m, 2H), 0.84 (t, $J = 7.6$ Hz, 3H); ^{13}C NMR (101 MHz, $CDCl_3$) δ 155.2, 83.4, 79.2, 43.3, 34.2, 32.9, 28.6, 25.1, 10.0; ^{11}B NMR (161 MHz, $CDCl_3$): δ 34.6; HRMS (ESI) *m/z*: $[M+H]^+$ calcd for $C_{18}H_{34}BNO_4H$ 340.2659; found 340.2647.

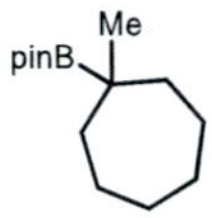

4,4,5,5-Tetramethyl-2-(1-methylcycloheptyl)-1,3,2-dioxaborolane (S20a)

Alkylboronic pinacol ester **S20a** was synthesized via a previously reported procedure [28]. Purification by flash column chromatography (silica gel, hexanes:CH_2Cl_2 = 8:1) afforded **S20a** (0.80 g, 67%) as a colorless oil; $R_f = 0.6$ (hexanes:CH_2Cl_2 = 3:1); ^{1}H NMR (400 MHz, $CDCl_3$) δ 1.84–1.74 (m, 2H), 1.62–1.37 (m, 8H), 1.22–1.14 (s, 14H), 0.90 (s, 3H); ^{13}C NMR (101 MHz, $CDCl_3$) δ 82.8, 38.7, 29.4, 26.6, 25.1, 24.8; ^{11}B NMR (161 MHz, $CDCl_3$): δ 35.0; HRMS (ESI) *m/z*: $[M+Na]^+$ calcd for $C_{14}H_{27}BO_2Na$ 261.2002; found 261.2002.

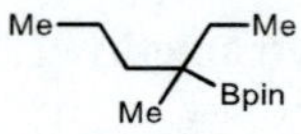

4,4,5,5-Tetramethyl-2-(3-methylhexan-3-yl)-1,3,2-dioxaborolane (S21a)
Alkylboronic pinacol ester **S21a** was synthesized via a previously reported procedure [28]. Purification by flash column chromatography (silica gel, hexanes:CH_2Cl_2 = 8:1) afforded **S21a** (0.48 g, 42%) as a colorless oil; R_f = 0.6 (hexanes:CH_2Cl_2 = 3:1); ^{1}H NMR (400 MHz, $CDCl_3$) δ 1.45–1.13 (m, 18H), 0.91–0.77 (m, 9H); ^{13}C NMR (101 MHz, $CDCl_3$) δ 83.0, 41.8, 31.8, 25.0, 21.0, 19.1, 15.3, 10.2; ^{11}B NMR (161 MHz, $CDCl_3$): δ 35.0; HRMS (ESI) *m/z*: $[M+Na]^+$ calcd for $C_{13}H_{27}BO_2Na$ 249.2002; found 249.2003.

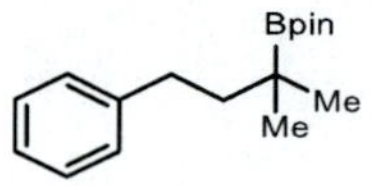

4,4,5,5-Tetramethyl-2-(2-methyl-4-phenylbutan-2-yl)-1,3,2-dioxaborolane (S22a)
Alkylboronic pinacol ester **S22a** was synthesized via a previously reported procedure [28]. Purification by flash column chromatography (silica gel, hexanes:CH_2Cl_2 = 7:3) afforded **S22a** (0.66 g, 48%) as a colorless oil; R_f = 0.5 (hexanes:CH_2Cl_2 = 3:1); ^{1}H NMR (400 MHz, $CDCl_3$) δ 7.27–7.23 (m, 2H), 7.20–7.11 (m, 3H), 2.59–2.50 (m, 2H), 1.63–1.53 (m, 2H), 1.24 (s, 12H), 0.99 (s, 6H); ^{13}C NMR (101 MHz, $CDCl_3$) δ 143.8, 128.5, 128.4, 125.6, 83.1, 43.7, 33.2, 24.9, 24.9; ^{11}B NMR (161 MHz, $CDCl_3$): δ 34.9; HRMS (ESI) *m/z*: $[M+Na]^+$ calcd for $C_{17}H_{27}BO_2Na$ 297.2002; found 297.2000.

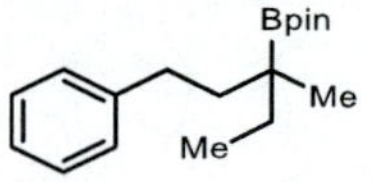

4,4,5,5-Tetramethyl-2-(3-methyl-1-phenylpentan-3-yl)-1,3,2-dioxaborolane (S24a)
Alkylboronic pinacol ester **S24a** was synthesized via a previously reported procedure [28]. Purification by flash column chromatography (silica gel, hexanes:CH_2Cl_2 = 7:3) afforded **S24a** (0.75 g, 52%) as a colorless oil; R_f = 0.5 (hexanes:CH_2Cl_2 = 3:1); ^{1}H NMR (400 MHz, $CDCl_3$) δ 7.26 (t, *J* = 7.5 Hz, 2H), 7.21–7.13 (m, 3H), 2.55 (m, 2H), 1.75–1.67 (m, 1H), 1.57–1.45 (m, 2H), 1.30–1.26 (m, 13H), 0.98 (s, 3H), 0.88 (t, *J* = 7.5 Hz, 3H); ^{13}C NMR (101 MHz, $CDCl_3$) δ 143.9, 128.5, 128.4, 125.6, 83.2, 41.5, 32.6, 31.5, 25.1, 25.0, 21.0, 10.2; ^{11}B NMR (161 MHz, $CDCl_3$): δ 35.0; HRMS (ESI) *m/z*: $[M+Na]^+$ calcd for $C_{18}H_{29}BO_2Na$ 311.2158; found 311.2156 (Fig. 4.3).

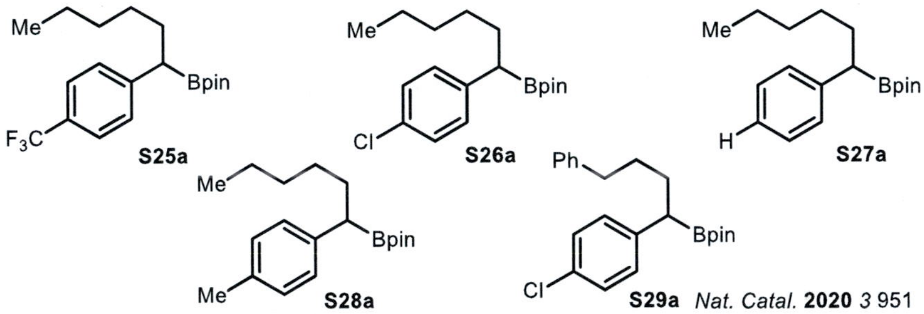

Fig. 4.3 List of alkylboronic pinacol esters for their preparation

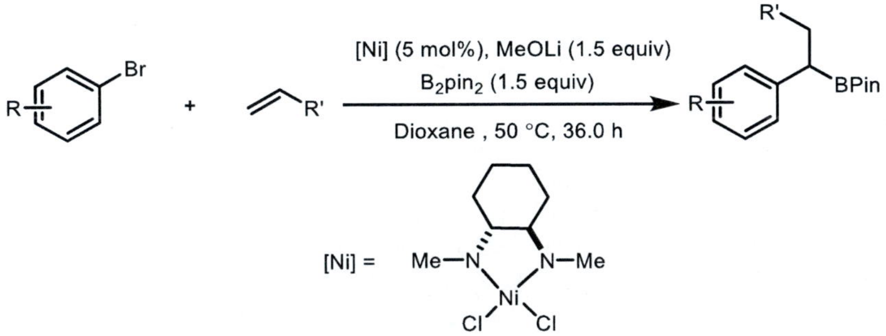

Alkylboronic pinacol esters (**S25a–S29a**) were synthesized via a previously reported procedure with slight modifications [29]. To a flask with a PTFE-coated stir bar were added Ni complex (5 mol%, 0.5 mmol), MeOLi (1.5 equiv, 15.0 mmol), and B_2Pin_2 (1.5 equiv, 15.0 mmol). After the flask was evacuated and backfilled with nitrogen (three cycles), dioxane (24 mL), aryl bromide (1.5 equiv, 15.0 mmol), and alkene (1.0 equiv, 10 mmol) were added to the flask, successively. Then, the reaction mixture was stirred at 50 °C for 36.0 h. The reaction mixture was filtered over a short pad of celite and the filter cake was washed with dichloromethane. The combined organic phase was concentrated *in vacuo* and purified by flash column chromatography (silica gel, hexanes:CH_2Cl_2 = 10:1) to afford the desired product.

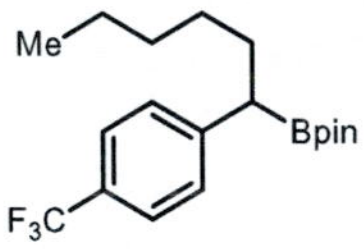

4,4,5,5-Tetramethyl-2-(1-(4-(trifluoromethyl)phenyl)hexyl)-1,3,2-dioxaborolane (S25a)

Purification by flash column chromatography (silica gel, hexanes:CH_2Cl_2 = 10:1) afforded **S25a** (2.57 g, 72%) as a colorless oil; R_f = 0.6 (hexanes: CH_2Cl_2 = 2:1);

^{1}H NMR (400 MHz, $CDCl_3$): δ 7.49 (d, J = 8.0 Hz, 2H), 7.30 (d, J = 8.0 Hz, 2H), 2.36 (t, J = 7.9 Hz, 1H), 1.92–1.78 (m, 1H), 1.70–1.60 (m, 1H), 1.34–1.14 (m, 18H), 0.89–0.81 (m, 3H); ^{13}C NMR (101 MHz, $CDCl_3$): δ 148.0, 128.69, 127.5 (q, J_{C-F} = 32.0 Hz), 125.3 (q, J_{C-F} = 3.8 Hz), 124.7 (q, J_{C-F} = 271.5 Hz), 83.6, 32.4, 31.9, 29.0, 24.8, 24.7, 22.6, 14.2; ^{19}F NMR (376 MHz, $CDCl_3$): δ −62.7; ^{11}B NMR (193 MHz, $CDCl_3$): δ 33.3; HRMS (ESI) *m/z*: $[M+Na]^+$ calcd for $C_{19}H_{28}BF_3O_2Na$ 379.2032; found 379.2031.

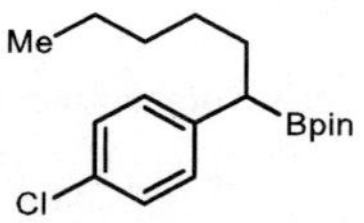

2-(1-(4-Chlorophenyl)hexyl)-4,4,5,5-tetramethyl-1,3,2-dioxaborolane (S26a)
Purification by flash column chromatography (silica gel, hexanes:CH_2Cl_2 = 10:1) afforded **S26a** (1.87 g, 58%) as a colorless oil; R_f = 0.6 (hexanes:CH_2Cl_2 = 2:1); ^{1}H NMR (500 MHz, $CDCl_3$): δ 7.21 (d, J = 8.3 Hz, 2H), 7.13 (d, J = 8.2 Hz, 2H), 2.27 (t, J = 7.9 Hz, 1H), 1.85–1.77 (m, 1H), 1.66–1.54 (m, 1H), 1.29–1.17 (m, 18H), 0.88–0.82 (m, 3H); ^{13}C NMR (126 MHz, $CDCl_3$): δ 142.2, 130.8, 129.8, 128.4, 83.5, 32.6, 31.9, 29.0, 24.8, 24.7, 22.7, 14.2; ^{11}B NMR (193 MHz, $CDCl_3$): δ 33.4; HRMS (ESI) *m/z*: $[M+Na]^+$ calcd for $C_{18}H_{28}BClO_2Na$ 345.1769; found 345.1767.

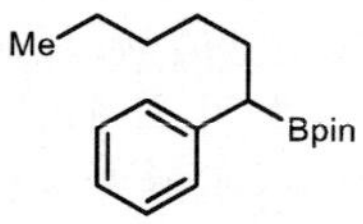

4,4,5,5-Tetramethyl-2-(1-phenylhexyl)-1,3,2-dioxaborolane (S27a)
Purification by flash column chromatography (silica gel, hexanes:CH_2Cl_2 = 10:1) afforded **S27a** (1.38 g, 48%) as a colorless oil; R_f = 0.6 (hexanes:CH_2Cl_2 = 2:1); ^{1}H NMR (400 MHz, $CDCl_3$): δ 7.30–7.14 (m, 4H), 7.11 (d, J = 7.1 Hz, 1H), 2.28 (t, J = 8.0 Hz, 1H), 1.91–1.75 (m, 1H), 1.69–1.57 (m, 1H), 1.33–1.11 (m, 18H), 0.90–0.78 (m, 3H); ^{13}C NMR (101 MHz, $CDCl_3$): δ 143.5, 128.3, 128.2, 125.0, 83.2, 32.5, 31.8, 28.9, 24.6, 24.5, 22.5, 14.0; ^{11}B NMR (193 MHz, $CDCl_3$): δ 33.1; HRMS (ESI) *m/z*: $[M+Na]^+$ calcd for $C_{18}H_{29}BO_2Na$ 311.2158; found 311.2157.

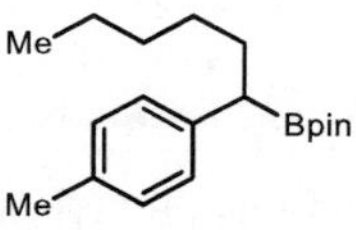

4,4,5,5-Tetramethyl-2-(1-(p-tolyl)hexyl)-1,3,2-dioxaborolane (S28a)
Purification by flash column chromatography (silica gel, hexanes:CH_2Cl_2 = 10:1) afforded **S28a** (1.36 g, 45%) as a colorless oil; $R_f = 0.6$ (hexanes:CH_2Cl_2 = 2:1); ^{1}H NMR (500 MHz, $CDCl_3$): δ 7.12 (d, *J* = 8.1 Hz, 2H), 7.07 (d, *J* = 7.8 Hz, 2H), 2.31 (s, 3H), 2.28 (t, *J* = 7.9 Hz, 1H), 1.88–1.79 (m, 1H), 1.69–1.58 (m, 1H), 1.31–1.21 (m, 18H), 0.90–0.85 (m, 3H); ^{13}C NMR (126 MHz, $CDCl_3$): δ 140.4, 134.4, 129.0, 128.3, 83.2, 32.9, 32.0, 29.1, 24.7, 24.7, 22.7, 21.1, 14.2; ^{11}B NMR (193 MHz, $CDCl_3$): δ 33.9; HRMS (ESI) *m/z*: $[M+Na]^+$ calcd for $C_{19}H_{31}BO_2Na$ 325.2315; found 325.2314.

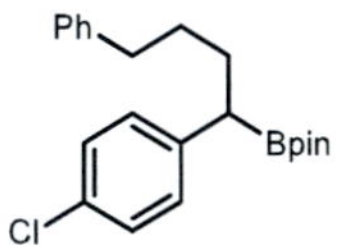

2-(1-(4-Chlorophenyl)-4-phenylbutyl)-4,4,5,5-tetramethyl-1,3,2-dioxaborolane (S29a)
Purification by flash column chromatography (silica gel, hexanes:CH_2Cl_2 = 10:1) afforded **S29a** (1.92 g, 52%) as a colorless oil; $R_f = 0.5$ (hexanes:CH_2Cl_2 = 2:1); ^{1}H NMR (400 MHz, $CDCl_3$): δ 7.34–7.04 (m, 9H), 2.68–2.50 (m, 2H), 2.30 (t, *J* = 7.9 Hz, 1H), 1.96–1.80 (m, 1H), 1.75–1.60 (m, 1H), 1.59–1.54 (m, 2H), 1.23–1.14 (m, 12H); ^{13}C NMR (101 MHz, $CDCl_3$): δ 142.6, 141.8, 131.0, 129.8, 128.5, 128.5, 128.4, 125.8, 83.6, 36.0, 32.3, 31.1, 24.8, 24.7; ^{11}B NMR (193 MHz, $CDCl_3$): δ 33.3; HRMS (ESI) *m/z*: $[M+Na]^+$ calcd for $C_{22}H_{28}BClO_2Na$ 393.1769; found 393.1766 (Fig. 4.4).

4.4.2.1 General Procedure for Conversion of Alkyl Pinacolboronate to Alkyl Trifluoroborate

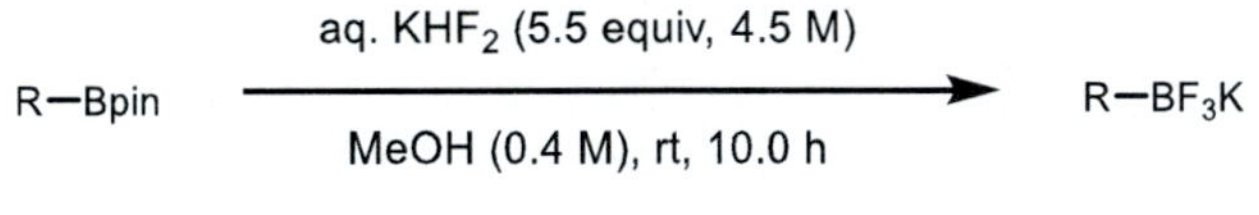

To a stirred solution of potassium hydrogen difluoride (4.5 M in H_2O, 5.5 equiv) at room temperature was added dropwise a solution of alkyl pinacolboronate **S1a–S29a** (0.4 M in MeOH, 1.0 equiv). The reaction mixture was then stirred at the same temperature for 10.0 h. After 10.0 h, the resulting suspension was concentrated under reduced pressure. H_2O was removed as a toluene azeotrope under reduced pressure.

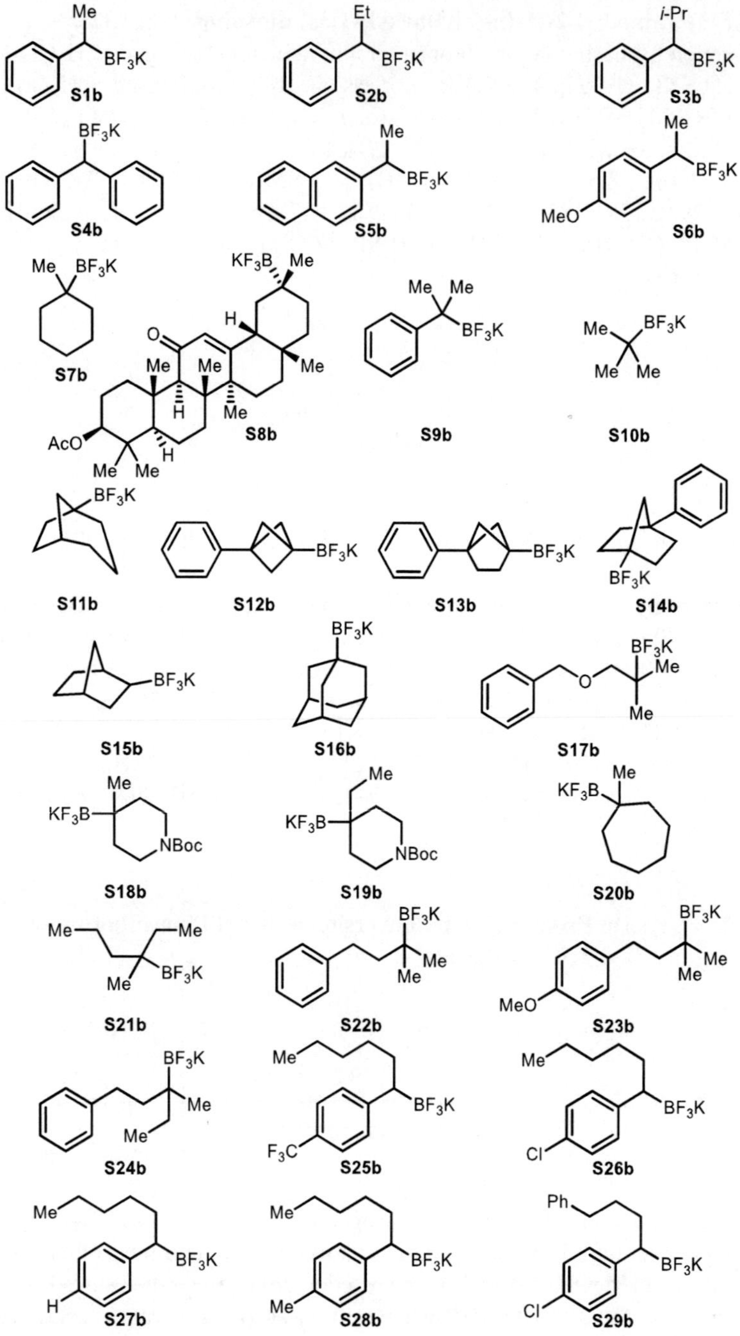

Fig. 4.4 List of alkyl trifluoroborate

The resultant solid was resuspended in hot acetone and filtered. The filtrate was then concentrated to a minimal volume, and the alkyl trifluoroborate was precipitated by the addition of cold hexanes. The mixture was filtered, washed with cold hexanes, and then dried on high vacuum overnight to yield the desired alkyl trifluoroborate (**S1b–S29b**) (Figs. 4.5, 4.6, 4.7, and 4.8).

Spectral Data Matched that Reported in the Literature: Potassium trifluoro(1-phenylethyl)borate (**S1b**) [22], potassium trifluoro(1-phenylpropyl)borate (**S2b**) [22], potassium trifluoro(2-methyl-1-phenylpropyl)borate (**S3b**) [22], potassium trifluoro(1-(4-methoxyphenyl)ethyl)borate (**S6b**) [14], potassium *tert*-butyltrifluoroborate (**S10b**) [30], potassium (bicyclo[2.2.1]heptan-2-yl)trifluoroborate (**S15b**) [31], potassium adamantan-1-yl)trifluoroborate (**S16b**) [32].

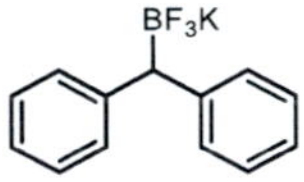

Potassium benzhydryltrifluoroborate (S4b)
White solid (2.15 g, 83%); m.p. 287–289 °C; ^{1}H NMR (400 MHz, DMSO-d_6): δ 7.20 (d, J = 7.0 Hz, 4H), 7.08 (t, J = 7.6 Hz, 4H), 6.93 (t, J = 7.3 Hz, 2H), 2.98 (s, 1H); ^{13}C NMR (101 MHz, DMSO-d_6): δ 148.9, 129.3, 127.0, 123.0; ^{11}B NMR (193 MHz, DMSO-d_6): δ 3.8; ^{19}F NMR (376 MHz, DMSO-d_6): δ −137.5; HRMS (ESI) m/z: [M]$^-$ calcd for $C_{13}H_{11}BF_3$ 235.0906; found 235.0913.

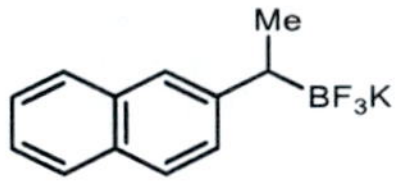

Potassium trifluoro(1-(naphthalen-2-yl)ethyl)borate (S5b)
White solid (1.12 g, 85%); m.p. 197–199 °C; ^{1}H NMR (400 MHz, DMSO-d_6): δ 7.73 (d, J = 8.1 Hz, 1H), 7.70 (d, J = 8.2 Hz, 1H), 7.59 (d, J = 8.4 Hz, 1H), 7.42 (s, 1H), 7.34 (d, J = 8.0 Hz, 2H), 7.28 (t, J = 7.3 Hz, 1H), 1.79 (s, 1H), 1.15 (d, J = 7.2 Hz, 3H); ^{13}C NMR (101 MHz, DMSO-d_6): δ 150.5, 133.4, 130.3, 128.9, 127.1, 126.7, 125.5, 124.8, 123.4, 123.2, 17.1; ^{11}B NMR (193 MHz, DMSO-d_6): δ 4.2; ^{19}F NMR (376 MHz, DMSO-d_6): δ −142.2; HRMS (ESI) m/z: [M]$^-$ calcd for $C_{12}H_{11}BF_3$ 223.0906; found 223.0913.

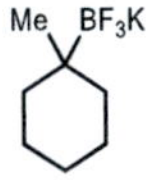

Potassium trifluoro(1-methylcyclohexyl)borate (S7b)
White solid (1.62 g, 69%); decomposed at 350 °C; ^{1}H NMR (500 MHz, DMSO-d_6): δ 1.40–1.23 (m, 7H), 1.22–1.10 (m, 1H), 0.94–0.82 (m, 2H), 0.60 (s, 3H); ^{13}C NMR (126 MHz, DMSO-d_6): δ 34.6, 27.4, 22.5, 22.0; ^{11}B NMR (193 MHz, DMSO-d_6): δ 4.8; ^{19}F NMR (376 MHz, DMSO-d_6): δ −148.3; HRMS (ESI) *m*/*z*: [M]$^-$ calcd for $C_7H_{13}BF_3$ 165.1062; found 165.1069.

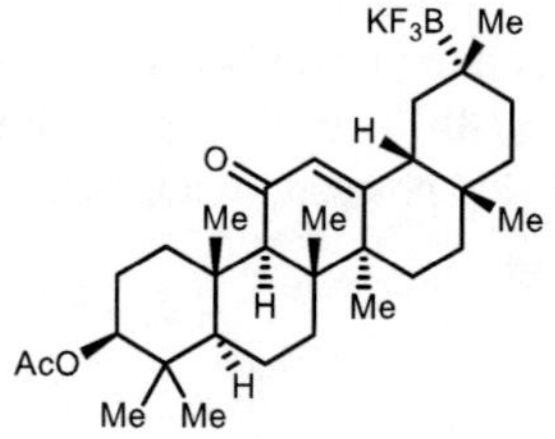

Potassium ((2*S*,4a*R*,6a*S*,6b*R*,8a*R*,10*S*,12a*S*,12b*R*,14b*R*)-10-acetoxy-2,4a,6a,6b,9,9,12a-heptamethyl-13-oxo-1,2,3,4,4a,5,6,6a,6b,7,8,8a,9,10,11,12,12a,12b,13,14b-icosahydropicen-2-yl)trifluoroborate (S8b)
White solid (1.26 g, 92%); m.p. 305–307 °C; ^{1}H NMR (400 MHz, DMSO-d_6): δ 5.36 (s, 1H), 4.43 (dd, *J* = 11.8, 4.5 Hz, 1H), 2.70–2.59 (m, 1H), 2.39 (s, 1H), 2.11–1.97 (m, 2H), 2.00 (s, 3H), 1.84 (t, *J* = 13.4 Hz, 1H), 1.79–1.55 (m, 3H), 1.55–1.28 (m, 6H), 1.33 (s, 3H), 1.15–0.99 (m, 3H), 1.06 (s, 3H), 1.04 (s, 3H), 0.94–0.74 (m, 4H), 0.82 (s, 6H), 0.76 (s, 3H), 0.64 (s, 3H); ^{13}C NMR (101 MHz, acetone-d_6): δ 199.6, 173.9, 170.7, 127.7, 80.8, 62.1, 55.4, 47.1, 45.9, 44.1, 40.0, 39.2, 38.5, 37.6, 35.9, 33.4, 33.2, 29.7, 28.3, 28.0, 27.3, 27.2, 24.2, 23.9, 21.1, 19.1, 18.0, 18.0, 17.0, 16.8; ^{11}B NMR (193 MHz, acetone-d_6): δ 5.3; ^{19}F NMR (376 MHz, acetone-d_6): δ −154.9; HRMS (ESI) *m*/*z*: [M]$^-$ calcd for $C_{31}H_{47}BF_3O_3$ 535.3570; found 535.3576.

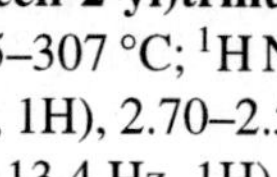

Potassium trifluoro(2-phenylpropan-2-yl)borate (S9b)
White solid (2.24 g, 97%); m.p. 181–183 °C; ^{1}H NMR (400 MHz, DMSO-d_6): δ 7.24 (d, *J* = 7.5 Hz, 2H), 7.07 (t, *J* = 7.6 Hz, 2H), 6.88 (t, *J* = 7.2 Hz, 1H), 1.03 (s, 6H); ^{13}C NMR (101 MHz, DMSO-d_6): δ 155.4, 126.5, 126.4, 121.9, 26.1; ^{11}B NMR (193 MHz, DMSO-d_6): δ 4.3; ^{19}F NMR (376 MHz, DMSO-d_6): δ −147.1; HRMS (ESI) *m*/*z*: [M]$^-$ calcd for $C_9H_{11}BF_3$ 187.0906; found 187.0905.

Potassium bicyclo[3.2.1]octan-1-yltrifluoroborate (S11b)
White solid (0.62 g, 92%); m.p. >350 °C; ^{1}H NMR (400 MHz, DMSO-d_6): δ 1.97 (s, 1H), 1.49–0.96 (m, 12H); ^{13}C NMR (101 MHz, DMSO-d_6): δ 41.7, 35.8, 34.2, 33.4, 31.2, 30.3, 20.1; ^{11}B NMR (193 MHz, DMSO-d_6): δ 4.8; ^{19}F NMR (376 MHz, DMSO-d_6): δ −147.2; HRMS (ESI) m*m/z*: [M]$^-$ calcd for $C_8H_{13}BF_3$ 177.1062; found 177.1069.

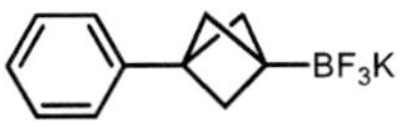

Potassium trifluoro(3-phenylbicyclo[1.1.1]pentan-1-yl)borate (S12b)
White solid (0.32 g, 59%); m.p. 310–312 °C; ^{1}H NMR (500 MHz, DMSO-d_6): δ 7.26–7.20 (m, 2H), 7.15–7.09 (m, 3H), 1.66–1.60 (m, 6H); ^{13}C NMR (126 MHz, DMSO-d_6): δ 144.0, 127.8, 125.5, 125.4, 51.1, 44.1; ^{11}B NMR (161 MHz, DMSO-d_6): δ 2.7; ^{19}F NMR (471 MHz, DMSO-d_6): δ −143.1; HRMS (ESI) *m/z*: [M]$^-$ calcd for $C_{11}H_{11}BF_3$ 211.0911; found 211.0913.

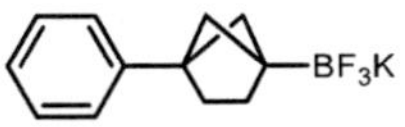

Potassium trifluoro(4-phenylbicyclo[2.1.1]hexan-1-yl)borate (S13b)
White solid (0.24 g, 61%); m.p. 332–334 °C; ^{1}H NMR (500 MHz, DMSO-d_6): δ 7.23 (t, J = 7.5 Hz, 2H), 7.17 (d, J = 7.1 Hz, 2H), 7.10 (t, J = 7.2 Hz, 1H), 1.71–1.64 (m, 2H), 1.57–1.50 (m, 2H), 1.41–1.34 (m, 2H), 1.14–1.08 (m, 2H); ^{13}C NMR (126 MHz, DMSO-d_6): δ 146.5, 127.8, 125.6, 125.0, 53.5, 44.4, 35.4, 31.3; ^{11}B NMR (161 MHz, DMSO-d_6): δ 4.0; ^{19}F NMR (471 MHz, DMSO-d_6): δ −143.8; HRMS (ESI) *m/z*: [M]$^-$ calcd for $C_{12}H_{13}BF_3$ 225.1068; found 225.1069.

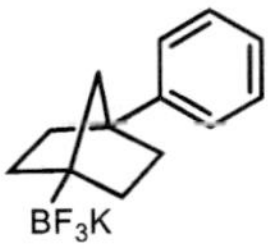

Potassium trifluoro(4-phenylbicyclo[2.2.1]heptan-1-yl)borate (S14b)
White solid (0.52 g, 68%); decomposed at 327 °C; ^{1}H NMR (400 MHz, DMSO-d_6): δ 7.29–7.17 (m, 4H), 7.09 (t, J = 6.9 Hz, 1H), 1.66–1.43 (m, 6H), 1.34–1.26 (m, 2H), 1.15–1.06 (m, 2H); ^{13}C NMR (126 MHz, DMSO-d_6): δ 148.3, 127.8, 126.4, 124.9, 52.9, 45.8, 38.9, 33.6; ^{11}B NMR (161 MHz, DMSO-d_6): δ 5.1; ^{19}F NMR (471 MHz, DMSO-d_6): δ −144.4; HRMS (ESI) m/z: [M]$^-$ calcd for $C_{13}H_{15}BF_3$ 239.1224; found 239.1225.

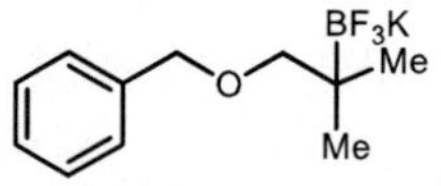

Potassium (1-(benzyloxy)-2-methylpropan-2-yl)trifluoroborate (S17b)
White solid (0.34 g, 70%); m.p. 122–124 °C; ^{1}H NMR (400 MHz, DMSO-d_6): δ 7.35–7.19 (m, 5H), 4.37 (s, 2H), 3.20 (s, 2H), 0.71 (s, 6H); ^{13}C NMR (101 MHz, DMSO-d_6): δ 140.3, 128.0, 126.9, 126.8, 80.1, 72.0, 22.0; ^{11}B NMR (193 MHz, DMSO-d_6): δ 4.4; ^{19}F NMR (376 MHz, DMSO-d_6): δ −48.9; HRMS (ESI) m/z: [M]$^-$ calcd for $C_{11}H_{15}BF_3O$ 231.1168; found 231.1171.

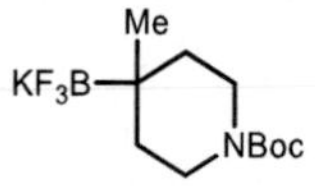

Potassium (1-(*tert*-butoxycarbonyl)-4-methylpiperidin-4-yl)trifluoroborate (S18b)
White solid (0.59 g, 65%); m.p. 271–273 °C; ^{1}H NMR (400 MHz, DMSO-d_6): δ 3.35–3.24 (m, 2H), 3.22–3.08 (m, 2H), 1.48–1.40 (m, 2H), 1.36 (s, 9H), 0.87–0.75 (m, 2H), 0.62 (s, 3H); ^{13}C NMR (126 MHz, DMSO-d_6): δ 154.2, 77.5, 41.1, 34.6, 34.4, 28.2, 23.0; ^{11}B NMR (161 MHz, DMSO-d_6): δ 4.8; ^{19}F NMR (471 MHz, DMSO-d_6): δ −148.4; HRMS (ESI) m/z: [M]$^-$ calcd for $C_{11}H_{20}BF_3NO_2$ 266.1545; found 266.1547.

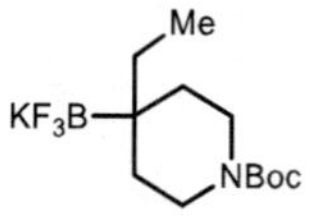

Potassium (1-(*tert*-butoxycarbonyl)-4-ethylpiperidin-4-yl)trifluoroborate (S19b)
White solid (0.46 g, 48%); m.p. 277–279 °C; ^{1}H NMR (400 MHz, DMSO-d_6): δ 3.29–3.16 (m, 4H), 1.47–1.39 (m, 2H), 1.36 (s, 9H), 1.13–1.06 (m, 2H), 0.92–0.81 (m, 2H), 0.73 (t, J = 7.5 Hz, 3H); ^{13}C NMR (126 MHz, DMSO-d_6): δ 154.3, 77.4, 41.3, 32.7, 32.3, 29.8, 28.3, 9.8; ^{11}B NMR (161 MHz, DMSO-d_6): δ 5.0; ^{19}F NMR (471 MHz, DMSO-d_6): δ −143.1; HRMS (ESI) *m/z*: [M]$^-$ calcd for $C_{12}H_{22}BF_3NO_2$ 280.1701; found 280.1701.

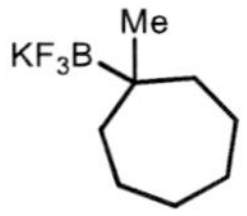

Potassium trifluoro(1-methylcycloheptyl)borate (S20b)
White solid (0.45 g, 98%); m.p. >350 °C; ^{1}H NMR (500 MHz, DMSO-d_6): δ 1.58–1.48 (m, 2H), 1.47–1.23 (m, 8H), 1.00–0.90 (m, 2H), 0.54 (s, 3H); ^{13}C NMR (101 MHz, DMSO-d_6): δ 37.3, 31.4, 25.4, 23.6; ^{11}B NMR (161 MHz, DMSO-d_6): δ 5.4; ^{19}F NMR (471 MHz, DMSO-d_6): δ −150.3; HRMS (ESI) *m/z*: [M]$^-$ calcd for $C_8H_{15}BF_3$ 179.1224; found 179.1225.

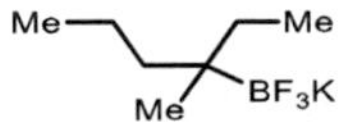

Potassium trifluoro(3-methylhexan-3-yl)borate (S21b)
White solid (0.14 g, 34%); m.p. 295–297 °C; ^{1}H NMR (400 MHz, DMSO-d_6): δ 1.23–1.13 (m, 2H), 1.11–1.01 (m, 2H), 0.99–0.92 (m, 2H), 0.75 (t, J = 7.3 Hz, 3H), 0.68 (t, J = 7.5 Hz, 3H), 0.52 (s, 3H); ^{13}C NMR (126 MHz, DMSO-d_6): δ 40.8, 30.0, 23.1, 17.5, 15.9, 9.5; ^{11}B NMR (161 MHz, DMSO-d_6): δ 5.1; ^{19}F NMR (471 MHz, DMSO-d_6): δ −145.3; HRMS (ESI) *m/z*: [M]$^-$ calcd for $C_7H_{15}BF_3$ 167.1224; found 167.1226.

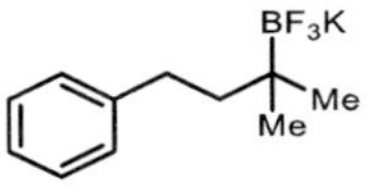

Potassium trifluoro(2-methyl-4-phenylbutan-2-yl)borate (S22b)
White solid (0.71 g, 85%); m.p. 207–209 °C; ^{1}H NMR (500 MHz, acetone-d_6): δ 7.19 (t, J = 7.5 Hz, 2H), 7.15 (d, J = 7.3 Hz, 2H), 7.06 (t, J = 7.1 Hz, 1H), 2.63–2.55 (m, 2H), 1.47–1.37 (m, 2H), 0.78 (s, 6H); ^{13}C NMR (126 MHz, acetone-d_6): δ 147.0, 129.2, 128.8, 125.5, 44.9, 32.7, 25.3; ^{11}B NMR (161 MHz, acetone-d_6): δ 5.8; ^{19}F NMR (471 MHz, acetone-d_6): δ −151.5; HRMS (ESI) m/z: [M]$^-$ calcd for $C_{11}H_{15}BF_3$ 215.1224; found 215.1226.

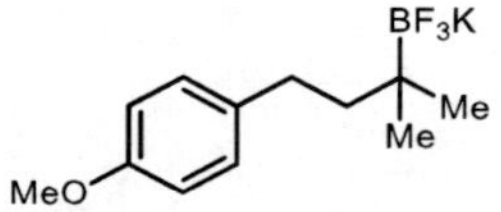

Potassium trifluoro(4-(4-methoxyphenyl)-2-methylbutan-2-yl)borate (S23b)
White solid (0.21 g, 73%); m.p. 181–183 °C; ^{1}H NMR (500 MHz, DMSO-d_6): δ 7.00 (d, J = 8.4 Hz, 2H), 6.77 (d, J = 8.4 Hz, 2H), 3.69 (s, 3H), 2.44–2.36 (m, 2H), 1.28–1.15 (m, 2H), 0.65 (s, 6H); ^{13}C NMR (126 MHz, DMSO-d_6): δ 156.7, 137.5, 128.8, 113.4, 54.9, 44.3, 30.5, 25.4; ^{11}B NMR (161 MHz, DMSO-d_6): δ 5.2; ^{19}F NMR (471 MHz, DMSO-d_6): δ −148.6 HRMS (ESI) m/z: [M]$^-$ calcd for $C_{12}H_{17}BF_3O$ 245.1330; found 245.1328.

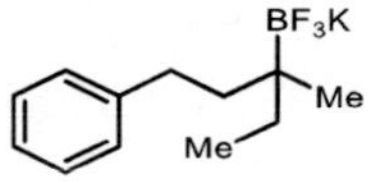

Potassium trifluoro(3-methyl-1-phenylpentan-3-yl)borate (S24b)
White solid (0.63 g, 78%); m.p. 140–142 °C; ^{1}H NMR (500 MHz, DMSO-d_6): δ 7.21 (t, J = 7.4 Hz, 2H), 7.13–7.04 (m, 3H), 2.50–2.39 (m, 2H), 1.32–1.23 (m, 2H), 1.22–1.09 (m, 2H), 0.75 (t, J = 7.4 Hz, 3H), 0.62 (s, 3H); ^{13}C NMR (126 MHz, DMSO-d_6): δ 145.8, 128.1, 128.0, 124.7, 40.9, 31.4, 29.8, 22.9, 9.4; ^{11}B NMR (161 MHz, DMSO-d_6): δ 5.1; ^{19}F NMR (471 MHz, DMSO-d_6): δ −145.4 HRMS (ESI) m/z: [M]$^-$ calcd for $C_{12}H_{17}BF_3$ 229.1381; found 229.1381.

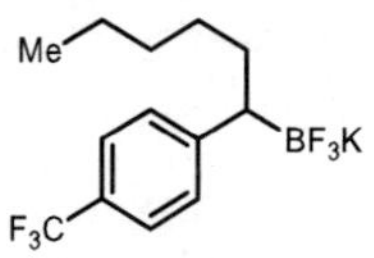

Potassium trifluoro(1-(4-(trifluoromethyl)phenyl)hexyl)borate (S25b)
White solid (1.92 g, 79%); m.p. 125–127 °C; ^{1}H NMR (400 MHz, DMSO-d_6): δ 7.38 (d, J = 7.9 Hz, 2H), 7.18 (d, J = 7.8 Hz, 2H), 1.73–1.43 (m, 3H), 1.30–0.95 (m, 6H), 0.88–0.69 (m, 3H); ^{13}C NMR (101 MHz, DMSO-d_6): δ 156.4, 128.5, 125.2 (q, $J_{C–F}$ = 271 Hz), 123.6 (q, $J_{C–F}$ = 4 Hz), 123.1 (q, $J_{C–F}$ = 31 Hz), 31.9, 31.0, 29.0, 22.2, 14.1; ^{11}B NMR (193 MHz, DMSO-d_6): δ 3.7; ^{19}F NMR (376 MHz, DMSO-d_6): δ −59.6, −141.6; HRMS (ESI) m/z: [M]$^-$ calcd for $C_{13}H_{16}BF_6$ 297.1249; found 297.1255.

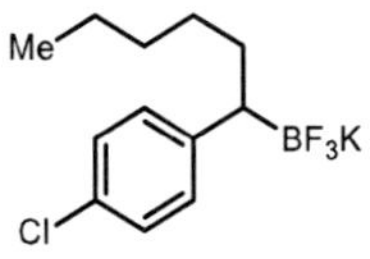

Potassium (1-(4-chlorophenyl)hexyl)trifluoroborate (S26b)
White solid (1.49 g, 85%); m.p. 181–183 °C; ^{1}H NMR (400 MHz, DMSO-d_6): δ 7.07 (d, J = 8.3 Hz, 2H), 6.99 (d, J = 8.3 Hz, 2H), 1.65–1.52 (m, 1H), 1.49–1.35 (m, 2H), 1.27–0.95 (m, 6H), 0.79 (t, J = 6.7 Hz, 3H); ^{13}C NMR (126 MHz, DMSO-d_6): δ 149.9, 129.8, 126.6, 126.5, 31.8, 31.4, 28.9, 22.2, 14.1; ^{11}B NMR (193 MHz, acetone-d_6): δ 4.4; ^{19}F NMR (376 MHz, DMSO-d_6): δ −141.5; HRMS (ESI) m/z: [M]$^-$ calcd for $C_{12}H_{16}BClF_3$ 263.0986; found 263.0996.

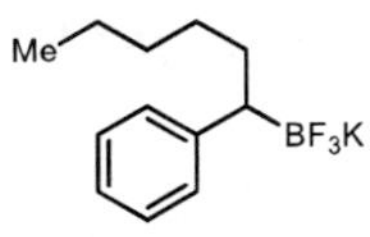

Potassium trifluoro(1-phenylhexyl)borate (S27b)
White solid (1.18 g, 92%); m.p. 193–195 °C; ^{1}H NMR (400 MHz, DMSO-d_6): δ 7.07–6.95 (m, 4H), 6.86 (t, J = 7.0 Hz, 1H), 1.65–1.52 (m, 1H), 1.51–1.35 (m, 2H), 1.25–0.97 (m, 6H), 0.78 (t, J = 6.7 Hz, 3H); ^{13}C NMR (126 MHz, DMSO-d_6): δ 150.9, 128.2, 126.6, 121.9, 31.8, 31.6, 28.9, 22.1, 14.0; ^{11}B NMR (193 MHz, acetone-d_6): δ 4.6; ^{19}F NMR (376 MHz, DMSO-d_6): δ −141.0; HRMS (ESI) m/z: [M]$^-$ calcd for $C_{12}H_{17}BF_3$ 229.1375; found 229.1370.

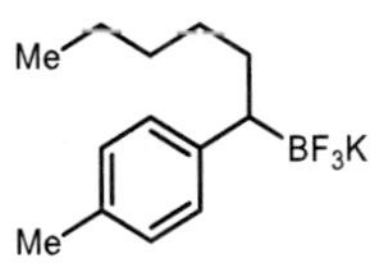

Potassium trifluoro(1-(p-tolyl)hexyl)borate (S28b)
White gum (0.84 g, 66%); ^{1}H NMR (500 MHz, DMSO-d_6): δ 6.87 (d, J = 8.0 Hz, 2H), 6.84 (d, J = 8.0 Hz, 2H), 2.18 (s, 3H), 1.60–1.50 (m, 1H), 1.45–1.31 (m, 2H), 1.26–0.95 (m, 6H), 0.78 (t, J = 7.0 Hz, 3H); ^{13}C NMR (126 MHz, DMSO-d_6): 147.6, 130.3, 128.1, 127.4, 31.9, 31.7, 29.0, 22.2, 20.6, 14.1; ^{11}B NMR (193 MHz, acetone-d_6): δ 4.6; ^{19}F NMR (376 MHz, DMSO-d_6): δ −141.1; HRMS (ESI) m/z: $[M]^-$ calcd for $C_{13}H_{19}BF_3$ 243.1532; found 243.1539.

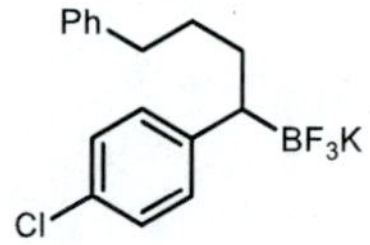

Potassium (1-(4-chlorophenyl)-4-phenylbutyl)trifluoroborate (S29b)
White solid (1.53 g, 84%); m.p. 114–116 °C; ^{1}H NMR (400 MHz, acetone-d_6): δ 7.30–7.00 (m, 9H), 2.63–2.40 (m, 2H), 1.88–1.76 (m, 1H), 1.75–1.13 (m, 4H); ^{13}C NMR (126 MHz, DMSO-d_6): 149.7, 143.0, 129.8, 128.2, 128.1, 126.6, 126.6, 125.3, 35.8, 31.5, 31.2; ^{11}B NMR (193 MHz, DMSO-d_6): δ 3.9; ^{19}F NMR (376 MHz, DMSO-d_6): δ −141.4; HRMS (ESI) m/z: $[M]^-$ calcd for $C_{16}H_{16}BClF_3$ 311.0986; found 311.0980.

4.4.3 Electrochemical Analyses for Reaction Design

All electrochemical measurements were performed with CHI 660 and 750 potentiostat (CH Instruments, TX, USA). The three-electrode system was used with Pt wire as the counter electrode and Ag/AgCl wire in 3.0 M NaCl as the reference electrode. The commercial glassy carbon disk electrode (3 mm inner diameter) was used as the working electrode. All the organic solutions were prepared with substrate (1 mM) in ^{n}Bu$_4$NPF$_6$ (0.1 M) of CH_2Cl_2 solution. Cyclic voltammetry was performed with a scan rate of 50 mV/s or 100 mV/s. For square wave voltammetry, a step potential of 5 mV, amplitude of 5 mV, and frequency of 10 Hz were used. Differential pulse voltammetry was implemented with a step potential of 4 mV, amplitude of 5 mV, pulse width of 50 ms, sampling width of 16.7 ms, and pulse period of 0.1 s. Note that the experiments above were conducted in N_2 atmosphere by purging.

Electrochemical measurements using thin-layer electroanalysis microchip (TEAM) were performed by following the literature [14]. Cyclic voltammetry was performed with the scan rate of 10 mV/s and n–t plot was obtained by the integrated

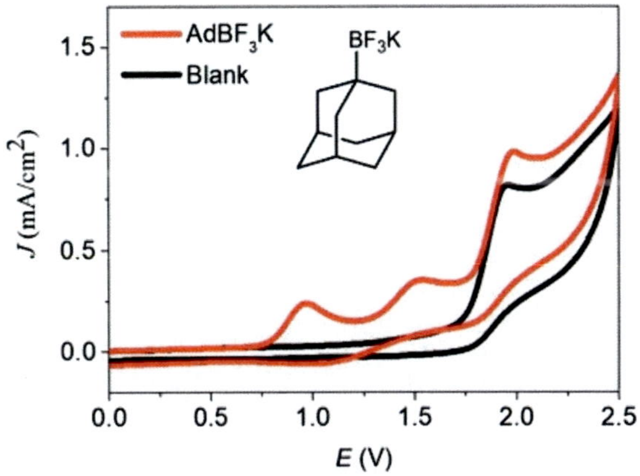

Fig. 4.5 Cyclic voltammogram of potassium (adamantan-1-yl)trifluoroborate S16b (1.0 mM) in CH_2Cl_2 solution with $^{n}Bu_4NPF_6$ (0.1 M) supporting electrolyte compared to that of blank (electrolyte solution). All scan rates of 100 mV/s

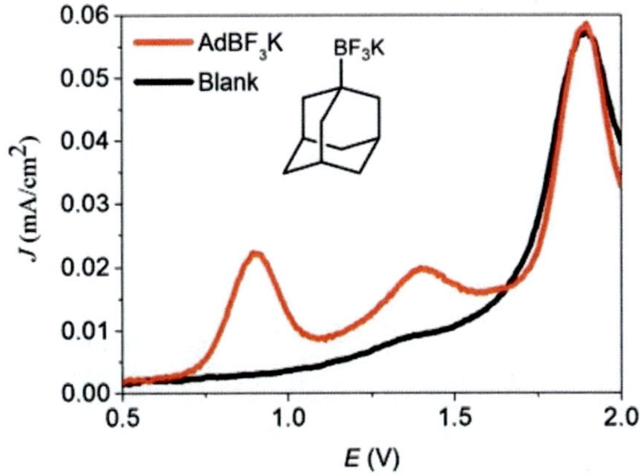

Fig. 4.6 Square wave voltammogram of potassium (adamantan-1-yl)trifluoroborate S16b (1.0 mM) in CH_2Cl_2 solution with $^{n}Bu_4NPF_6$ (0.1 M) supporting electrolyte compared to that of blank (electrolyte solution). The applied square wave is given as a 5-mV step potential, 5 mV amplitude, and 10 Hz frequency

data from the electrolysis using the chronoamperometry technique. Note that the electrode material of TEAM was indium tin oxide (ITO) of which overall electrochemical behavior was similar to the glassy carbon electrode [14].

4.4.3.1 Electrochemical Analyses of Alkyltrifluoroborate S16b

4.4.3.2 Electrochemical Analyses of Alkyltrifluoroborate S1b

See Figs. 4.9, 4.10, and 4.11.

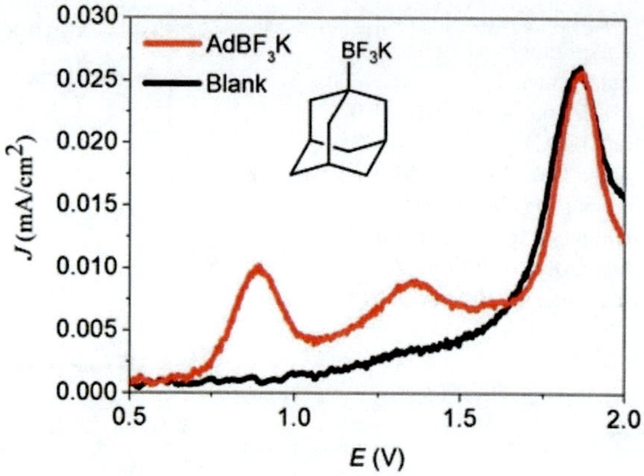

Fig. 4.7 Differential pulse voltammogram of potassium (adamantan-1-yl)trifluoroborate S16b (1.0 mM) in CH_2Cl_2 solution with $^nBu_4NPF_6$ (0.1 M) supporting electrolyte compared to that of blank (electrolyte solution). The applied differential pulse was given as a 4-mV step potential, 5 mV amplitude, 50 ms pulse width, 16.7 ms sample width, and 0.1 s pulse period

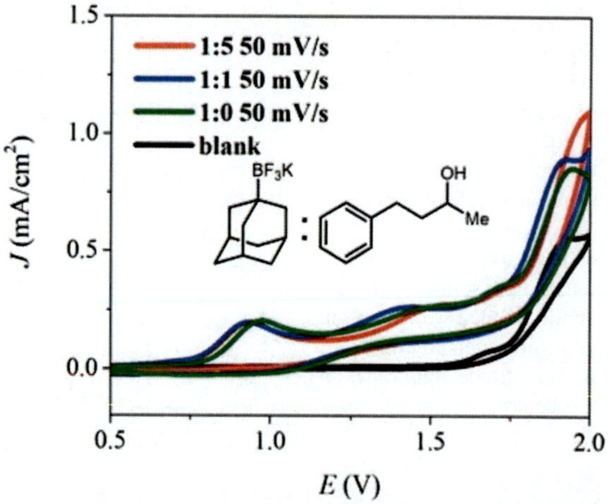

Fig. 4.8 Cyclic voltammograms of potassium (adamantan-1-yl)trifluoroborate S16b (1.0 mM) and alcohol (0.0 mM, 1.0 mM, or 5.0 mM) in CH_2Cl_2 solution with $^nBu_4NPF_6$ (0.1 M) supporting electrolyte compared to that of blank (electrolyte solution). All scan rates of 50 mV/s. No significant changes in oxidation potential as a function of alcohol concentration were observed

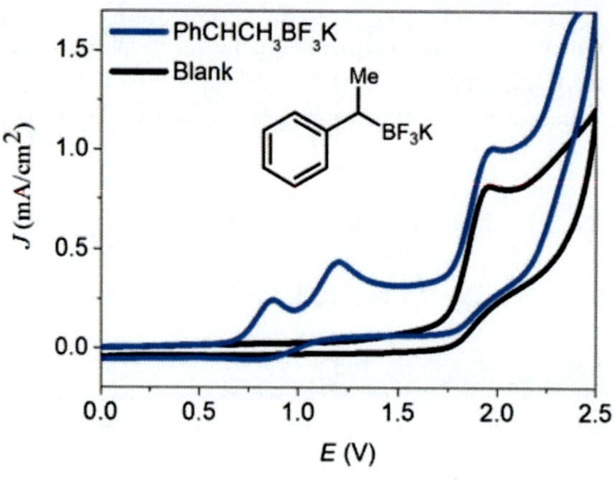

Fig. 4.9 Cyclic voltammogram of potassium trifluoro(1-phenylethyl)borate S1b (1.0 mM) in CH_2Cl_2 solution with $^nBu_4NPF_6$ (0.1 M) supporting electrolyte compared to that of blank (electrolyte solution). All scan rates of 100 mV/s

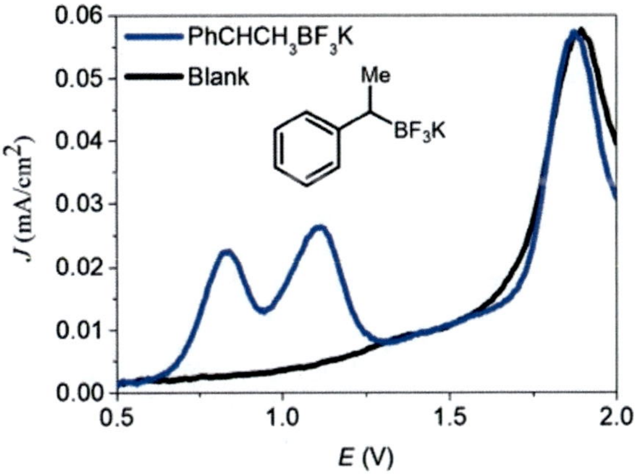

Fig. 4.10 Square wave voltammogram of potassium trifluoro(1-phenylethyl)borate S1b (1.0 mM) in CH_2Cl_2 solution with $^nBu_4NPF_6$ (0.1 M) supporting electrolyte compared to that of blank (electrolyte solution). The applied square wave was given as a 5-mV step potential, 5 mV amplitude, and 10 Hz frequency

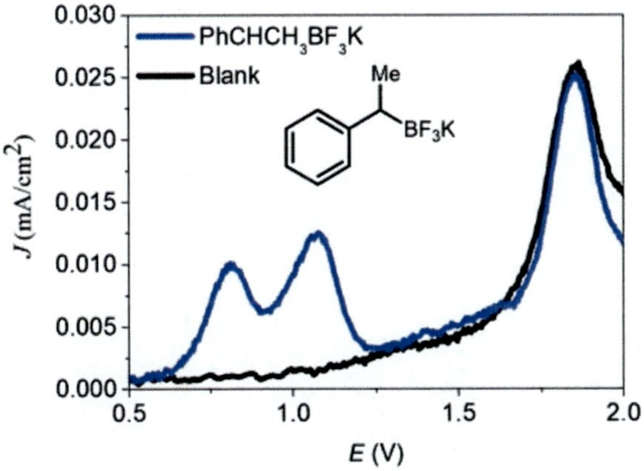

Fig. 4.11 Differential pulse voltammogram of potassium trifluoro(1-phenylethyl)borate S1b (1.0 mM) in CH_2Cl_2 solution with $^nBu_4NPF_6$ (0.1 M) supporting electrolyte compared to that of blank (electrolyte solution). The applied differential pulse was given as a 4-mV step potential, 5 mV amplitude, 50 ms pulse width, 16.7 ms sample width, and 0.1 s pulse period

4.4.3.3 Electrochemical Oxidation Study of Alkylboron Using TEAM

See Figs. 4.12 and 4.13.

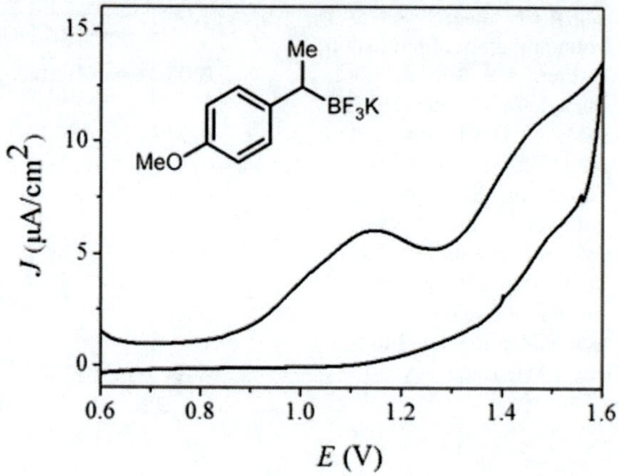

Fig. 4.12 Cyclic voltammogram of potassium trifluoro(1-(4-methoxyphenyl)ethyl)borate S6b (1.0 mM) in CH_2Cl_2 solution with $^nBu_4NPF_6$ (0.1 M) supporting electrolyte obtained by TEAM at the scan rate of 10 mV/s

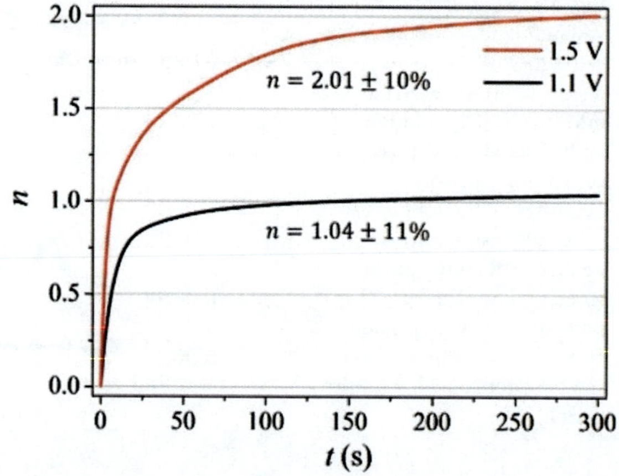

Fig. 4.13 N–t plots (Eapp = 1.1 and 1.5 V) of potassium trifluoro(1-(4-methoxyphenyl)ethyl)borate S6b (1.0 mM) in $^nBu_4NPF_6$ (0.1 M) of CH_2Cl_2 solution

4.4.3.4 Electrochemical Analyses of Nucleophiles

See Figs. 4.14, 4.15, 4.16, 4.17, and 4.18.

4.4.4 Reaction Optimization

See Table 4.3.

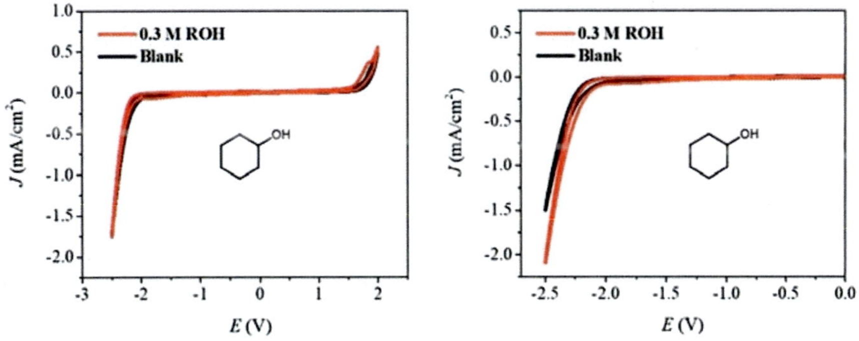

Fig. 4.14 Cyclic voltammograms of cyclohexanol (0.3 M) in CH_2Cl_2 solution with $^nBu_4NPF_6$ (0.1 M) supporting electrolyte compared to that of blank (electrolyte solution). All scan rates of 50 mV/s

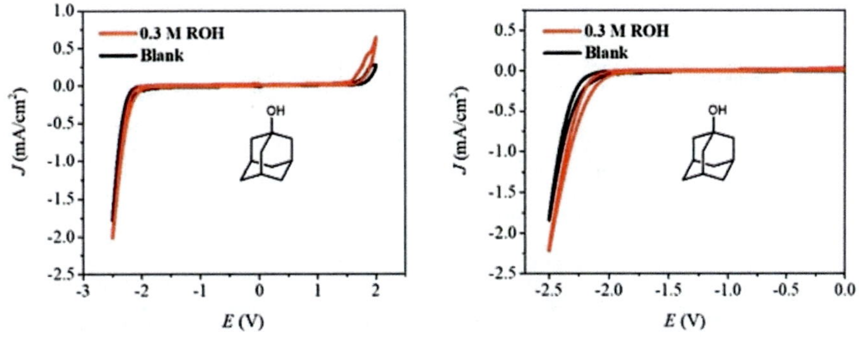

Fig. 4.15 Cyclic voltammograms of 1-adamantanol (0.3 M) in CH_2Cl_2 solution with $^nBu_4NPF_6$ (0.1 M) supporting electrolyte compared to that of blank (electrolyte solution). All scan rates of 50 mV/s

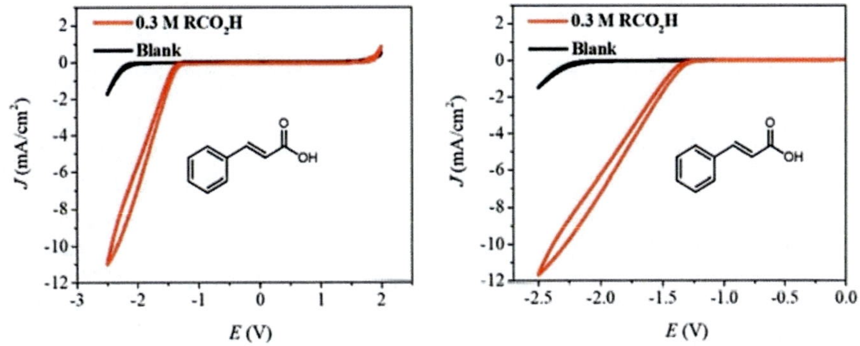

Fig. 4.16 Cyclic voltammograms of cinnamic acid (0.3 M) in CH_2Cl_2 solution with $^nBu_4NPF_6$ (0.1 M) supporting electrolyte compared to that of blank (electrolyte solution). All scan rates of 50 mV/s

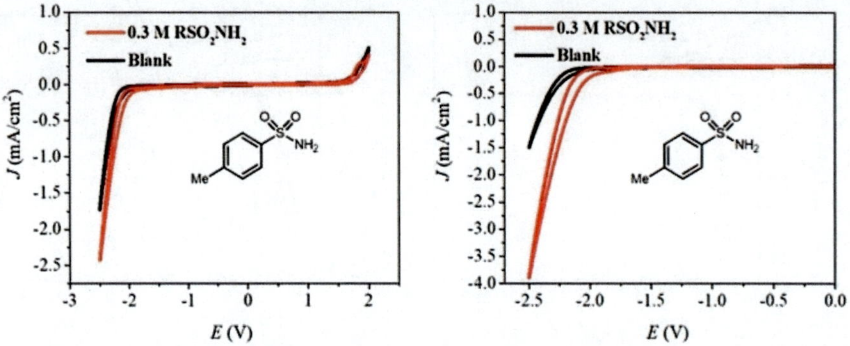

Fig. 4.17 Cyclic voltammograms of 4-methylbenzenesulfonamide (0.3 M) in CH_2Cl_2 solution with $^{n}Bu_4NPF_6$ (0.1 M) supporting electrolyte compared to that of blank (electrolyte solution). All scan rates of 50 mV/s

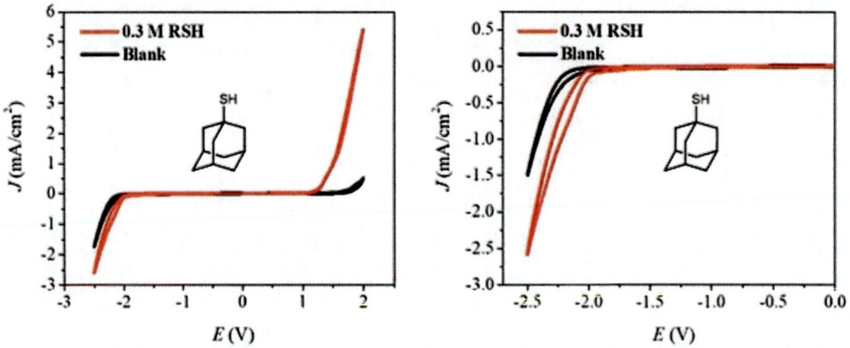

Fig. 4.18 Cyclic voltammograms of 1-Adamantanethiol (0.3 M) in CH_2Cl_2 solution with $^{n}Bu_4NPF_6$ (0.1 M) supporting electrolyte compared to that of blank (electrolyte solution). All scan rates of 50 mV/s

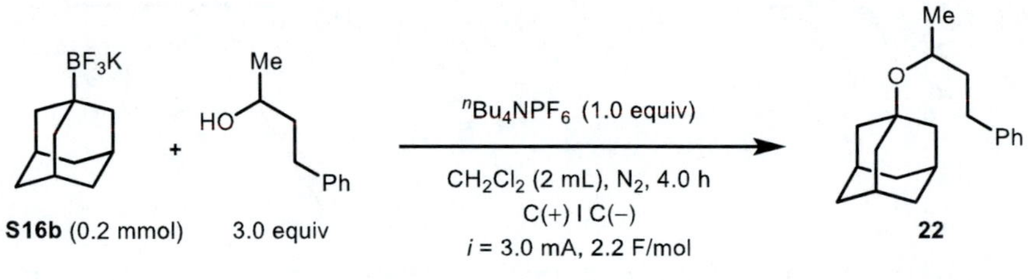

Table 4.3 Control experiments

Entry	Deviation	GC yield (%)
1	None	54
2	Air instead of N_2	8
3	No electrolyte	6
4	No electricity	–
5	AdBpin (**S16a**) instead of $AdBF_3K$ (**S16b**)	–

See Table 4.4.

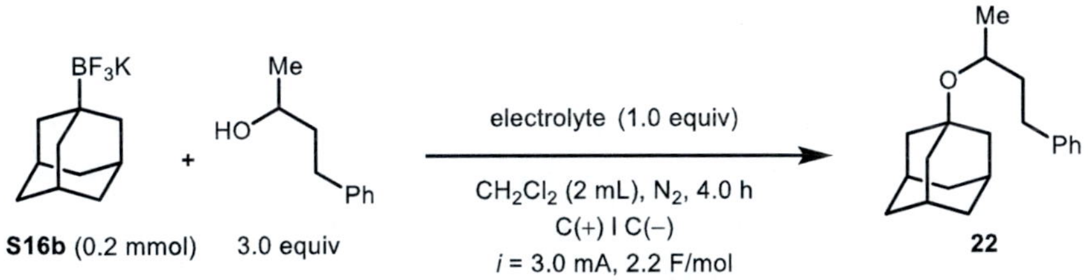

See Table 4.5.

Table 4.4 Evaluation of electrolytes

Entry	Electrolyte	GC yield (%)
1	$^{n}Bu_4NPF_6$	54
2	$^{n}Bu_4NClO_4$	36
3	$^{n}Bu_4NBF_4$	32
4	$^{n}Bu_4NBr$	4
5	$^{n}Bu_4NI$	5
6	LiCl	15
7	LiBr	4
8	$LiClO_4$	Trace
9	Me_3BnNCl	7
10	–	6

Table 4.5 Evaluation of solvents

Entry	Solvent	GC yield (%)
1	CH_2Cl_2	54
2	Toluene	–
3	DMF	–
4	DMSO	5
5	DCE	53
6	Trifluorotoluene	6
7	CH_3CN	5
8	1,4-Dioxane	Trace
9	1,4-Dichlorobutane	6

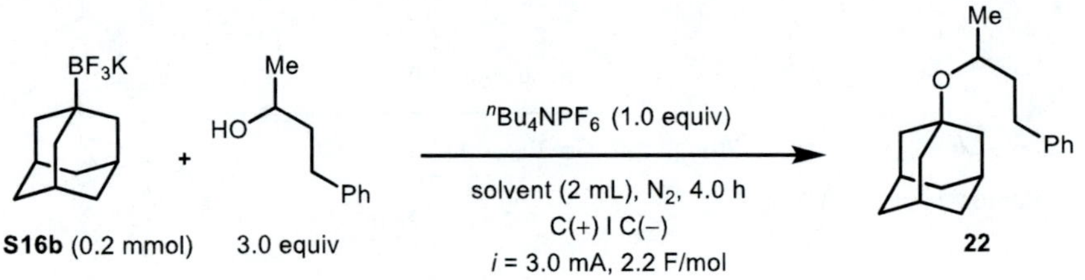

See Table 4.6.

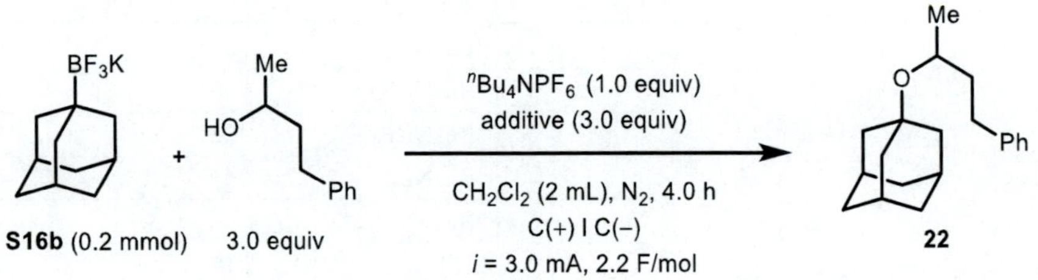

See Table 4.7.

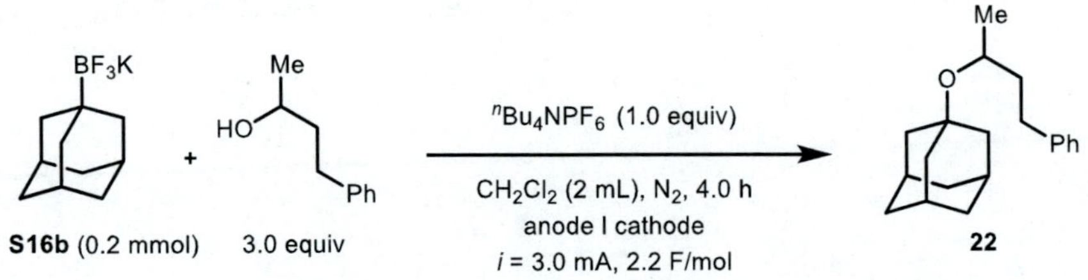

Table 4.6 Evaluation of additives

Entry	Additive	GC yield (%)
1	-	54
2	K_2CO_3	47
3	KOAc	27
4	2,4,6-Trimethylpyridine	Trace
5[a]	4 Å molecular sieve	38
6[b]	H_2O (1.4 equiv)	16

[a] 150 mg of 4 Å molecular sieve was added
[b] 5 μL of H_2O was added

Table 4.7 Evaluation of electrodes

Entry	Anode	Cathode	GC yield (%)
1	Graphite	Graphite	54
2	Platinum	Graphite	Trace
3	Stainless steel	Graphite	Trace
4	Nickel	Graphite	Trace
5	Glassy carbon	Graphite	17
6	Graphite	Platinum	31
7	Graphite	Stainless steel	15
8	Graphite	Nickel	37
9	Graphite	Glassy carbon	28

See Table 4.8.

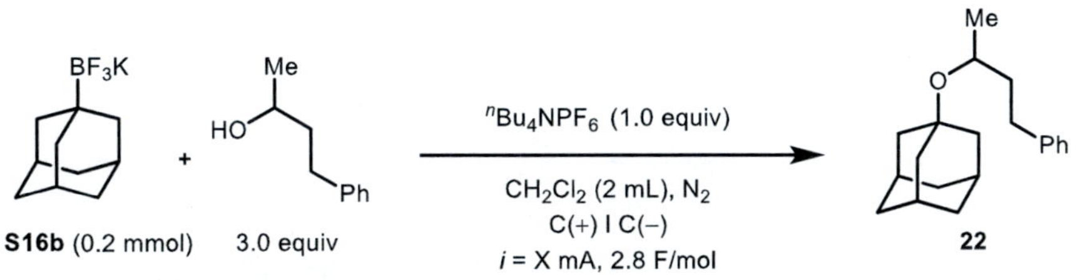

Table 4.8 Evaluation of current

Entry	Current (mA)	Time (min)	GC yield (%)
1	3.0	300	56
2	5.0	180	61
3	10.0	90	60

4.4.5 *General Procedure for Electrochemical C(sp^3)–Heteroatom Bond Formation*

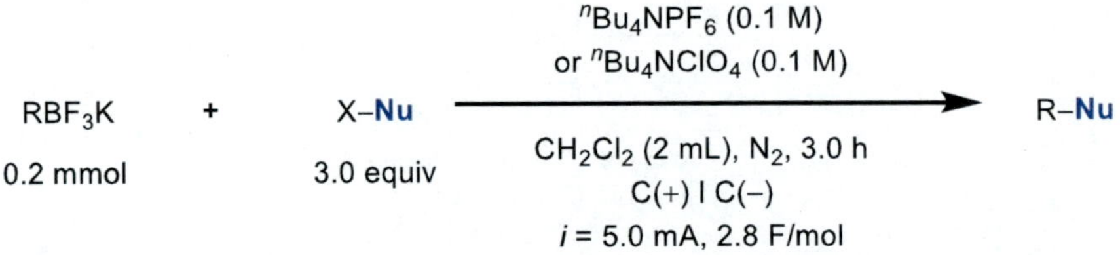

To an oven-dried undivided cell with a PTFE-coated stir bar were added alkyl trifluoroborate salt (0.2 mmol) and tetra-*n*-butylammonium hexafluorophosphate (0.2 mmol) or tetra-*n*-butylammonium perchlorate (0.2 mmol). The reaction cell was capped with a septa-lined teflon cap equipped with the anode (graphite) and the cathode (graphite). After the cell was evacuated and backfilled with nitrogen (three cycles), a balloon filled with nitrogen (1 atm) was connected to the reaction system. Subsequently, the coupling partner (0.6 mmol) dissolved in dichloromethane (2 mL) was added to the reaction cell. Then the reaction mixture was electrolyzed at a constant current of 5.0 mA for 3.0 h at room temperature. Upon completion of the reaction, the cap was removed, and electrodes were rinsed with dichloromethane (2 mL). The combined organic phase was transferred to a round bottom flask and concentrated *in vacuo*. The crude product was purified by flash column chromatography (silica gel, hexanes/CH_2Cl_2, or hexanes/EtOAc gradient elution) to afford the desired product (Fig. 4.19).

4.4.6 *Representative Larger Scale Reactions*

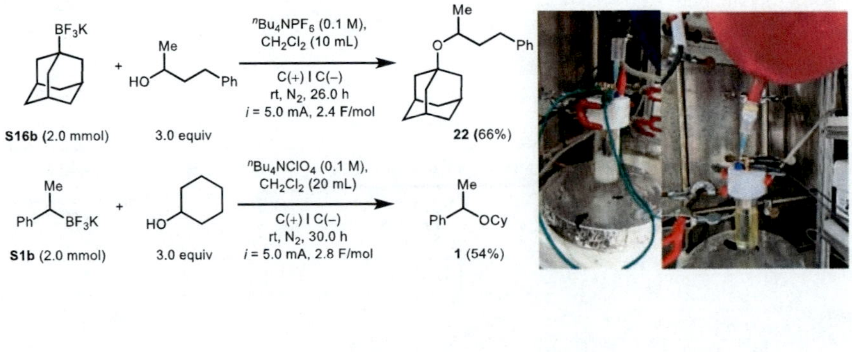

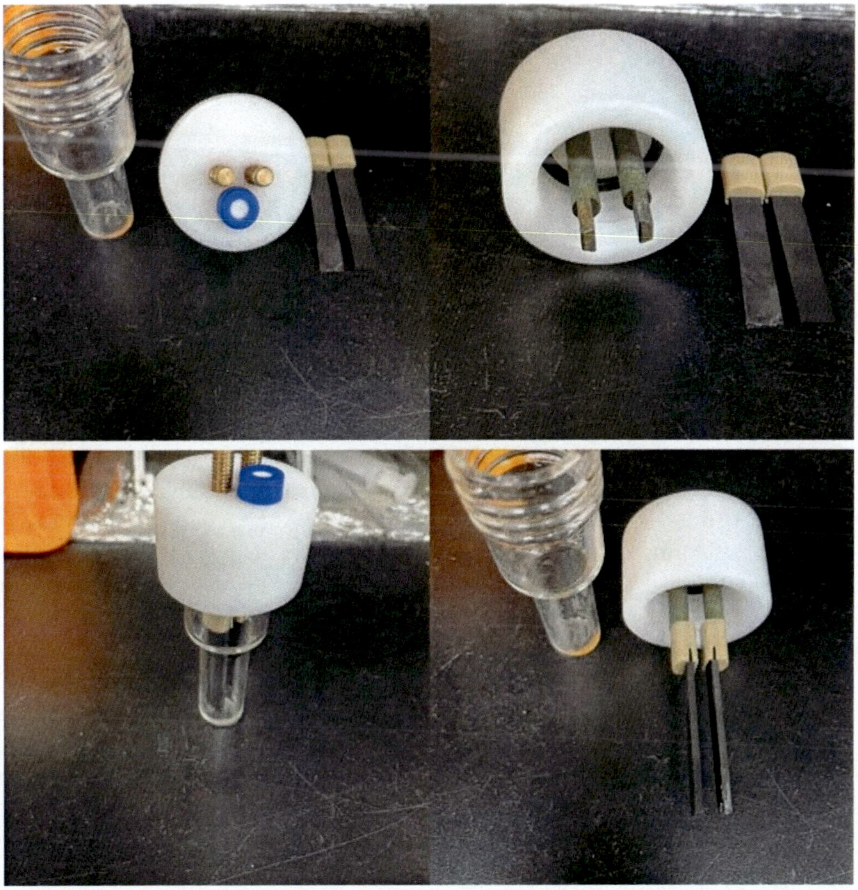

Fig. 4.19 Electrochemical reaction set-up

To an oven-dried undivided cell with a PTFE-coated stir bar were added alkyl trifluoroborate salt and tetra-*n*-butylammonium hexafluorophosphate or tetra-*n*-butylammonium perchlorate. The reaction cell was capped with a septa-lined teflon cap equipped with the anode (graphite) and the cathode (graphite). After the cell was evacuated and backfilled with nitrogen (three cycles), a balloon filled with nitrogen (1 atm) was connected to the reaction system. Subsequently, the coupling partner dissolved in dichloromethane was added to the reaction cell. Then the reaction mixture was electrolyzed at a constant current of 5.0 mA at room temperature. Upon completion of the reaction, the cap was removed, and electrodes were rinsed with dichloromethane (10 mL). The combined organic phase was transferred to a round bottom flask and concentrated *in vacuo*. The crude product was purified by flash column chromatography (silica gel, hexanes/CH_2Cl_2, or hexanes/EtOAc gradient elution) to afford the desired product.

4.4.7 Experimental Procedure for Electrochemical Bond Formation via In Situ Generation of Reactive Precursor

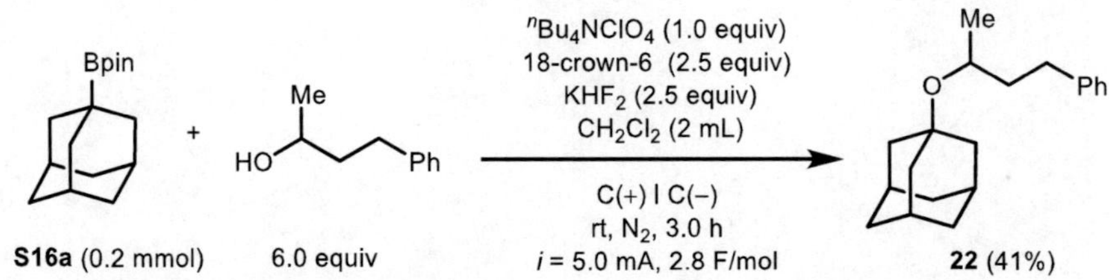

To an oven-dried undivided cell with a PTFE-coated stir bar were added alkyl pinacolboronate **S16a** (0.2 mmol), tetra-*n*-butylammonium perchlorate (1.0 equiv), 1,4,7,10,13,16-hexaoxacyclooctadecane (18-crown-6; 2.5 equiv), and potassium hydrogen difluoride (2.5 equiv). The reaction cell was capped with a septa-lined teflon cap equipped with the anode (graphite) and the cathode (graphite). After the cell was evacuated and backfilled with nitrogen (three cycles), a balloon filled with nitrogen (1 atm) was connected to the reaction system. Subsequently, Alcohol (6 equiv) dissolved in dichloromethane (2 mL) was added to the reaction cell. Then the reaction mixture was electrolyzed at a constant current of 5.0 mA for 3.0 h at room temperature. Upon completion of the reaction, the cap was removed, and electrodes were rinsed with dichloromethane (2 mL). The combined organic phase was transferred to a round bottom flask and concentrated *in vacuo*. The crude product was purified by flash column chromatography (silica gel, hexanes:CH_2Cl_2 = 7:3) to afford the desired product.

4.4.8 Divided Cell Experiments

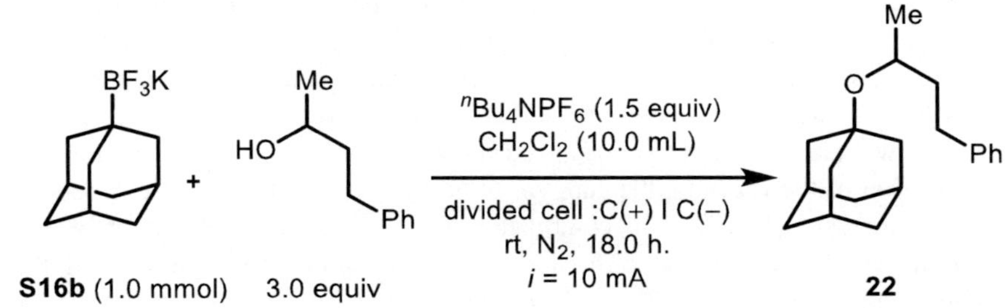

The electrolysis was carried out in an oven-dried H-type divided cell equipped with a PTFE-coated stir bar at each electrode. Graphite electrodes were used as the anode and the cathode, which were separated by an AMI-7001-30 membrane. To the anodic chamber was added alkyl trifluoroborate salt **S16b** (1.0 mmol) and tetra-*n*-butylammonium hexafluorophosphate (1.5 mmol) while to the cathodic chamber was added tetra-*n*-butylammonium hexafluorophosphate (1.5 mmol). After the reaction cell was evacuated and backfilled with nitrogen (three cycles), a balloon filled with nitrogen (1 atm) was connected to the electrolysis system. Subsequently, a solution of 4-phenyl-2-butanol (3.0 mmol) in CH_2Cl_2 (10 mL) was added to both the anodic chamber and cathodic chamber. Then the reaction mixture was electrolyzed at a constant current of 10.0 mA at room temperature for 18.0 h (entry1) (Table 4.9).

Table 4.9 Divided cell experiments

	Anodic chamber	Cathodic chamber	Membrane	Current (mA)	Time (h)	Anodic chamber **22** (%)[c]	Cathodic chamber **22** (%)[c]
1	**S16b**, alcohol $^nBu_4NPF_6$	Alcohol $^nBu_4NPF_6$	AMI-7001-30	10	18	8	–
2	**S16b**, alcohol $^nBu_4NPF_6$	Alcohol $^nBu_4NPF_6$	AMI-7001-30	3	20	3	–
3[a]	alcohol $^nBu_4NPF_6$	**S16b**, alcohol $^nBu_4NPF_6$	AMI-7001-30	3	20	–	–
4[b]	**S16b**, alcohol $^nBu_4NP_6$	Alcohol $^nBu_4NPF_6$	–	3	20	13	–

[a]Reaction conditions: alcohol (3.0 mmol), $^nBu_4NPF_6$ (1.5 mmol), CH_2Cl_2 (10.0 mL) [anodic chamber]. **S16b** (1.0 mmol), alcohol (3.0 mmol), $^nBu_4NPF_6$ (1.5 mmol), CH_2Cl_2 (10.0 mL) [cathodic chamber]
[b]Reaction conditions: **S16b** (1.0 mmol), alcohol (6.0 mmol), $^nBu_4NPF_6$ (3.0 mmol), CH_2Cl_2 (20.0 mL) [undivided cell]
[c]The reaction yield was determined by gas chromatography with mesitylene as the internal standard (Fig. 4.20)

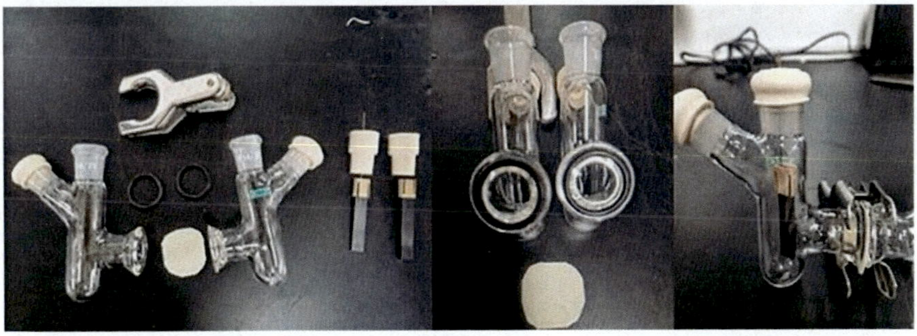

Fig. 4.20 Divided cell set-up

4.4.9 Constant Potential Electrolysis

All the electrochemical measurements were performed with CHI 660 and 750 potentiostat (CH Instruments, TX, USA). For constant potential electrolysis experiments, the reduced electric potential induces a significantly longer total reaction time in order to achieve the constant total charge transfer. Therefore, slightly reduced amount of the total charge (1.2 F/mol) was utilized to manage technical limitations associated with maintaining the inert reaction conditions, volatility of the reaction solvent, and the exponentially decreasing current of the system. The three-electrode system was used with graphite as the working and counter electrode and Ag/AgCl wire in 3.0 M NaCl as the reference electrode. The constant potential electrolysis was performed in various potentials and was terminated based on the total charge passed through the circuit, 1.2 F/mol. The reaction yield was determined by GC using mesitylene as an internal standard. All reactions were carried out under a nitrogen atmosphere. If a potential lower than 1.0 V was applied (vs Ag/AgCl in 3.0 M NaCl) to the reaction mixture, the reaction requires more than 15 days to reach the same total charge used in potential–yield diagram (Fig. 4.21).

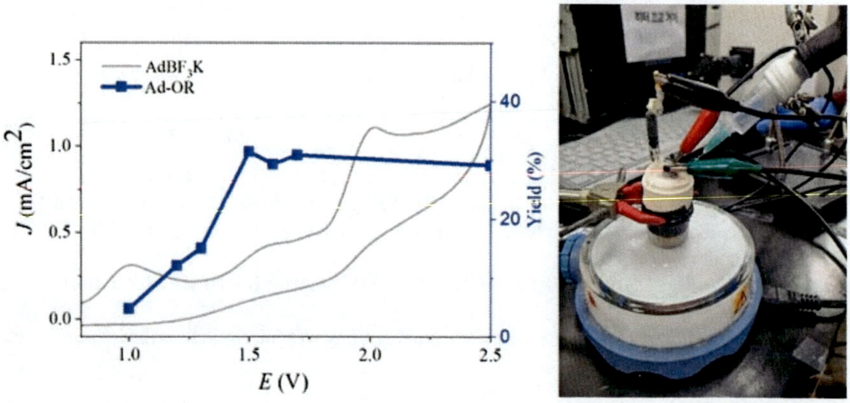

Fig. 4.21 (Left) Correlation plot between applied potentials and reaction yields. (Right) Three-electrode system with working electrode (green), counter electrode (red), and reference electrode (black). Ad-OR represents 22

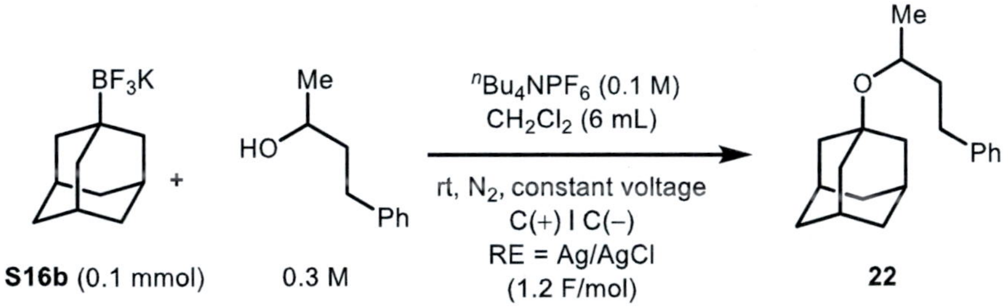

4.4.10 Mechanistic Probes and Kinetic Study

4.4.10.1 Substituent Effects in Substituted Aromatic Rings

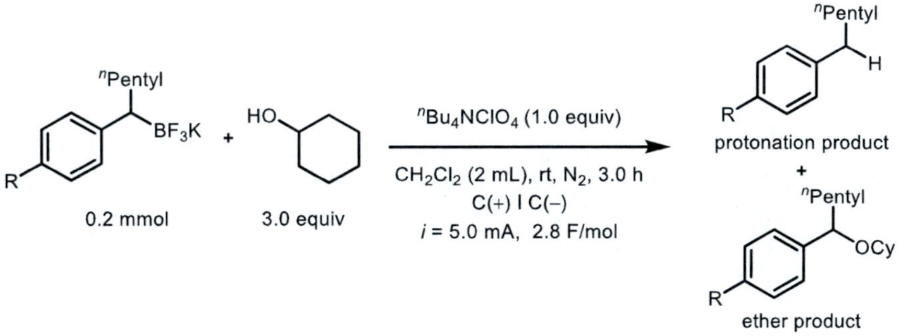

See Table 4.10.

The yield of protonation product was determined by ^{1}H NMR analysis using 1,1,2,2-tetrachloroethane as an internal standard. The yield of ether product was isolated yield.

See (Fig. 4.22, Table 4.11).

Table 4.10 Electronic effect of substituents

	Substituent constant [33, 34]	Protonation product (%)	Ether product (%)
R = CF_3	+0.54	40	18
R = Cl	+0.23	22	42
R = H	+0.00	13	58
R = Me	−0.17	Trace	59

Table 4.11 Chemical shifts of benzylic C–H of each observed compounds

	Protonation product	Ether product
R = CF_3	2.65 ppm (t, 7.8 Hz)	4.41 ppm (dd, 8.3, 4.9 Hz)
R = Cl	2.57 ppm (t, 7.8 Hz)	4.32 ppm (dd, 8.1, 5.2 Hz)
R = H	2.60 ppm (t, 7.8 Hz)	4.33 ppm (dd, 8.2, 5.1 Hz)
R = Me	–	4.32 ppm (dd, 8.2, 5.3 Hz)

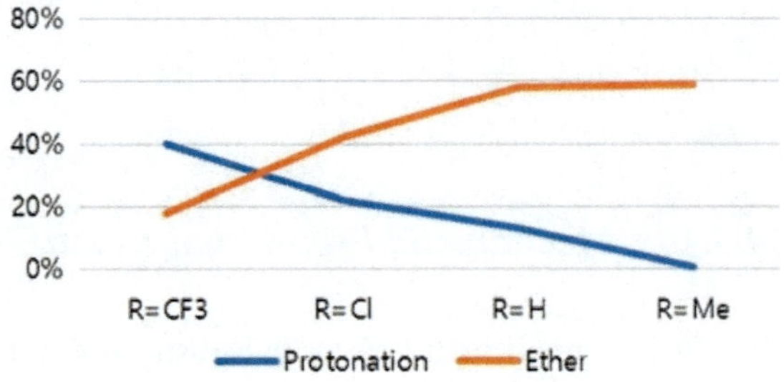

Fig. 4.22 Correlation plot between substituents and reaction yields

4.4.10.2 Evidence for the Carbocation Rearrangement Mechanism

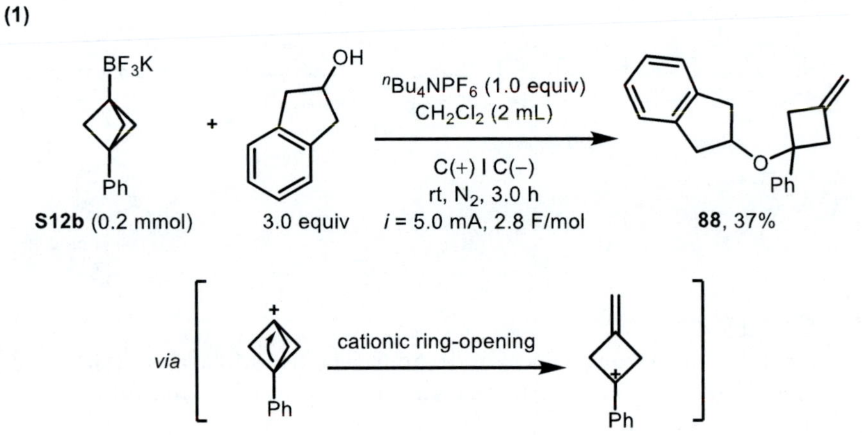

To an oven-dried undivided cell with a PTFE-coated stir bar was added alkyl trifluoroborate salt **S12b** (0.2 mmol) and tetra-*n*-butylammonium hexafluorophosphate (0.2 mmol). The reaction cell was capped with a septa-lined teflon cap equipped with the anode (graphite) and the cathode (graphite). After the cell was evacuated and backfilled with nitrogen (three cycles), a balloon filled with nitrogen (1 atm) was connected to the reaction system. Subsequently, 2,3-dihydro-1*H*-inden-2-ol (0.6 mmol) dissolved in dichloromethane (2 mL) was added to the reaction cell. Then the reaction mixture was electrolyzed at a constant current of 5.0 mA for 3.0 h at room temperature. Upon completion of the reaction, the cap was removed, and electrodes were rinsed with dichloromethane (2 mL). The combined organic phase was transferred to a round bottom flask and concentrated *in vacuo*. The crude product was purified by flash column chromatography (silica gel, hexanes/EtOAc gradient elution) to afford the desired product **88** (20.5 mg, 37%) as a colorless oil; R_f = 0.6 (hexanes:EtOAc = 10:1). Spectroscopic data are presented in the experimental procedures and characterization data section.

(2)

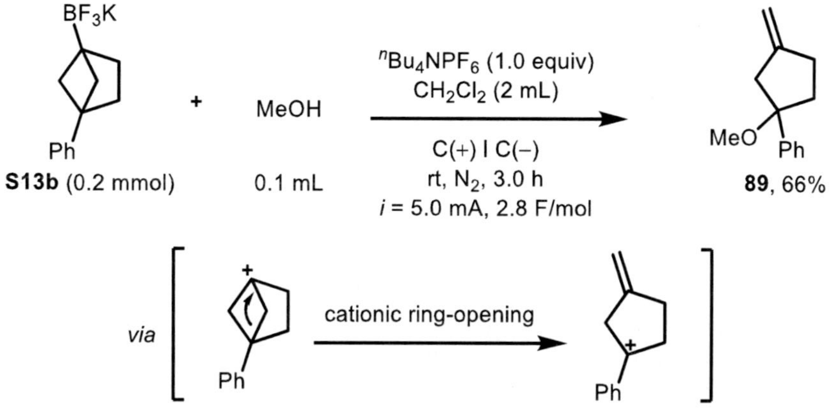

To an oven-dried undivided cell with a PTFE-coated stir bar was added alkyl trifluoroborate salt **S13b** (0.2 mmol) and tetra-*n*-butylammonium hexafluorophosphate (0.2 mmol). The reaction cell was capped with a septa-lined teflon cap equipped with the anode (graphite) and the cathode (graphite). After the cell was evacuated and backfilled with nitrogen (three cycles), a balloon filled with nitrogen (1 atm) was connected to the reaction system. Subsequently, methanol (0.1 mL) dissolved in dichloromethane (2 mL) was added to the reaction cell. Then the reaction mixture was electrolyzed at a constant current of 5.0 mA for 3.0 h at room temperature. Upon completion of the reaction, the cap was removed, and electrodes were rinsed with dichloromethane (2 mL). The combined organic phase was transferred to a round bottom flask and concentrated *in vacuo*. The crude product was purified by flash column chromatography (silica gel, hexanes/CH_2Cl_2 gradient elution) to afford the desired product **89** (25.0 mg, 66%) as a colorless oil; R_f = 0.5 (hexanes:EtOAc = 10:1). Spectroscopic data are presented in the experimental procedures and characterization data section.

(3)

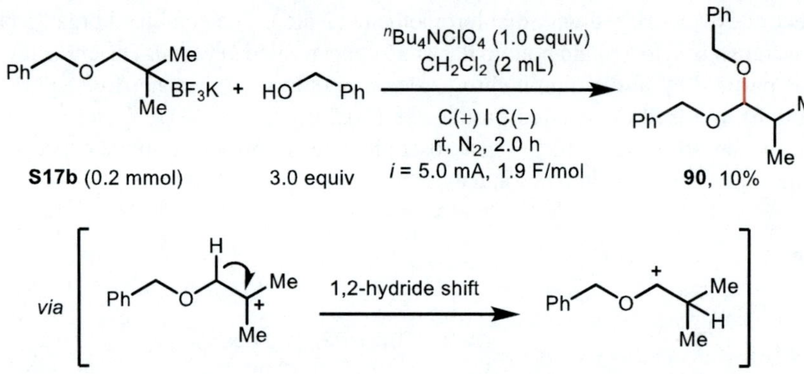

To an oven-dried undivided cell with a PTFE-coated stir bar were added alkyl trifluoroborate salt **S17b** (0.2 mmol) and tetra-*n*-butylammonium perchlorate (0.2 mmol). The reaction cell was capped with a septa-lined teflon cap equipped with the anode (graphite) and the cathode (graphite). After the cell was evacuated and backfilled with nitrogen (three cycles), a balloon filled with nitrogen (1 atm) was connected to the reaction system. Subsequently, benzyl alcohol (0.6 mmol) dissolved in dichloromethane (2 mL) was added to the reaction cell. Then the reaction mixture was electrolyzed at a constant current of 5.0 mA for 2.0 h at room temperature. Upon completion of the reaction, the cap was removed, and electrodes were rinsed with dichloromethane (2 mL). The combined organic phase was transferred to a round bottom flask and concentrated *in vacuo*. The crude product was purified by flash column chromatography (silica gel, hexanes/CH_2Cl_2 gradient elution) to afford the desired product **90** (5.5 mg, 10%) as a colorless oil; $R_f = 0.5$ (hexanes:EtOAc = 1:1). Spectroscopic data matches with previously reported data [35].

4.4.10.3 Reaction Profiles for the Kinetic Analysis

All electrochemical measurements were performed with dual display potentiostat (DJS-292B and DJS-292C, China). The reaction yield was determined by GC using *n*-octane as an internal standard. All reactions were carried out under a nitrogen atmosphere (Figs. 4.23, 4.24, 4.25, 4.26, 4.27, 4.28, 4.29, and 4.30).

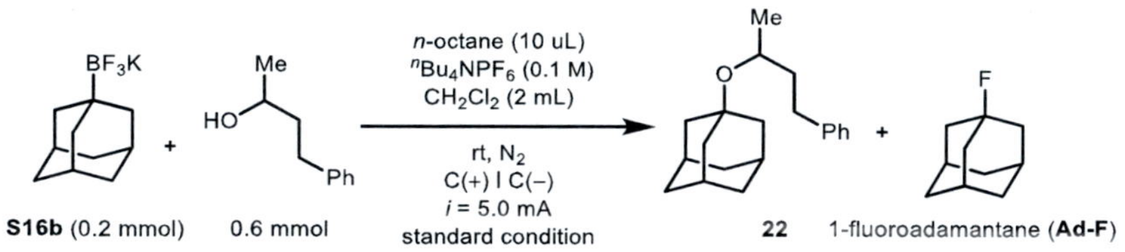

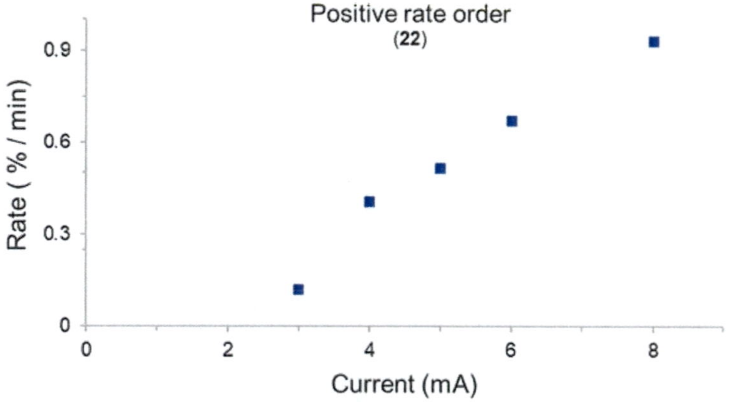

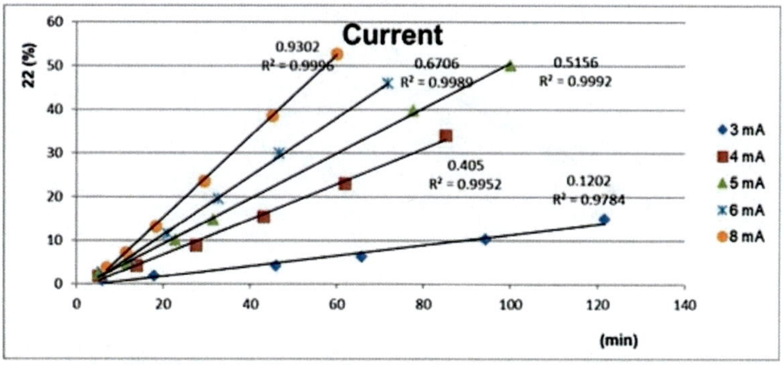

Fig. 4.23 (Left) Dependence of reaction rate, formation of 22, on the current of the cell. The result shows near first-order kinetics with respect to the applied current. Due to the unchanged ratio of fluoride and alcohol nucleophile, only the alcohol product was analyzed. (Right) Plot of yield (22) in a given time period under varying currents. The yields of 22 were calculated based on the amount of S16b

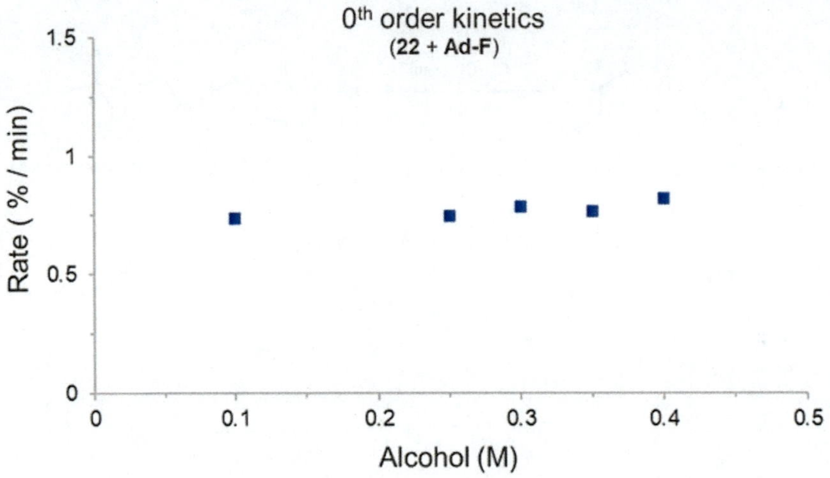

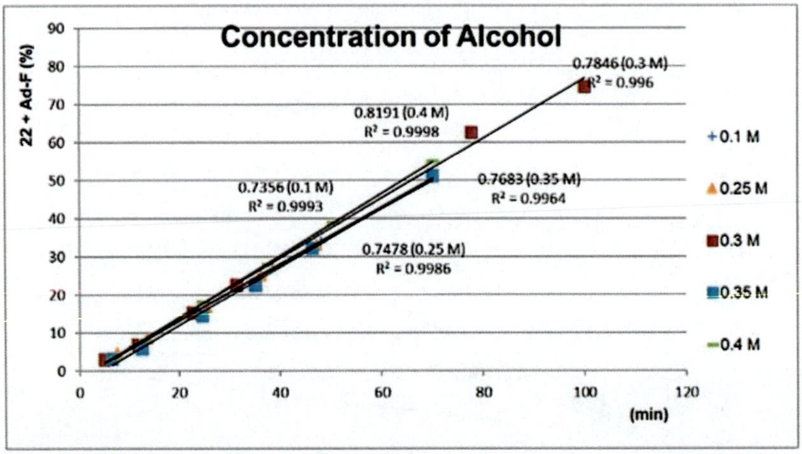

Fig. 4.24 (Left) Dependence of reaction rate, formation of 22 and Ad-F, on the concentration of alcohol. The result shows zero-order kinetics with respect to the alcohol concentration. (Right) Plot of yield (cation-trapping products 22 and Ad-F) in a given period of time under varying concentrations of alcohol. The yields of 22 and Ad-F were calculated based on the amount of S16b

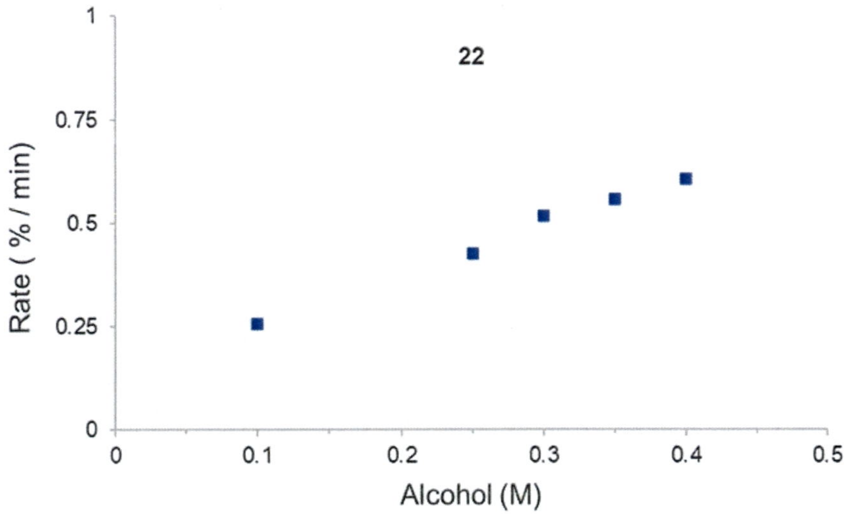

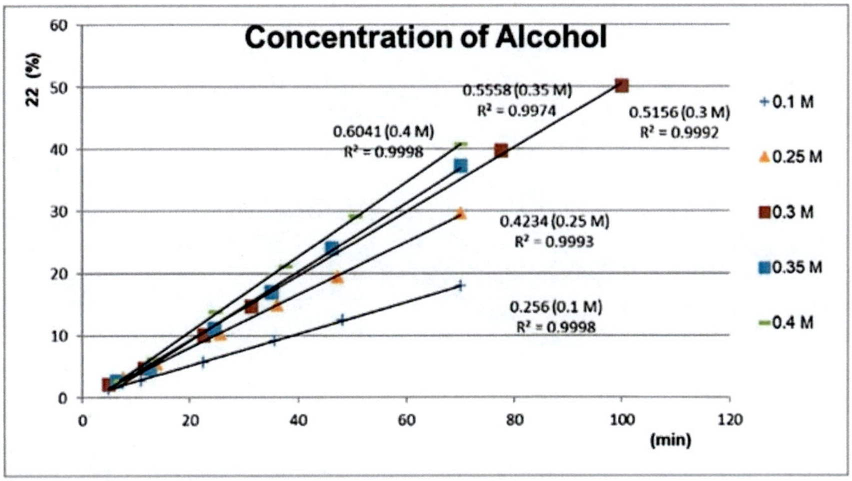

Fig. 4.25 (Left) Dependence of reaction rate, formation of 22, on the concentration of alcohol. (Right) Plot of yield (22) in a given period of time under varying concentrations of alcohol. The yields of 22 were calculated based on the amount of S16b

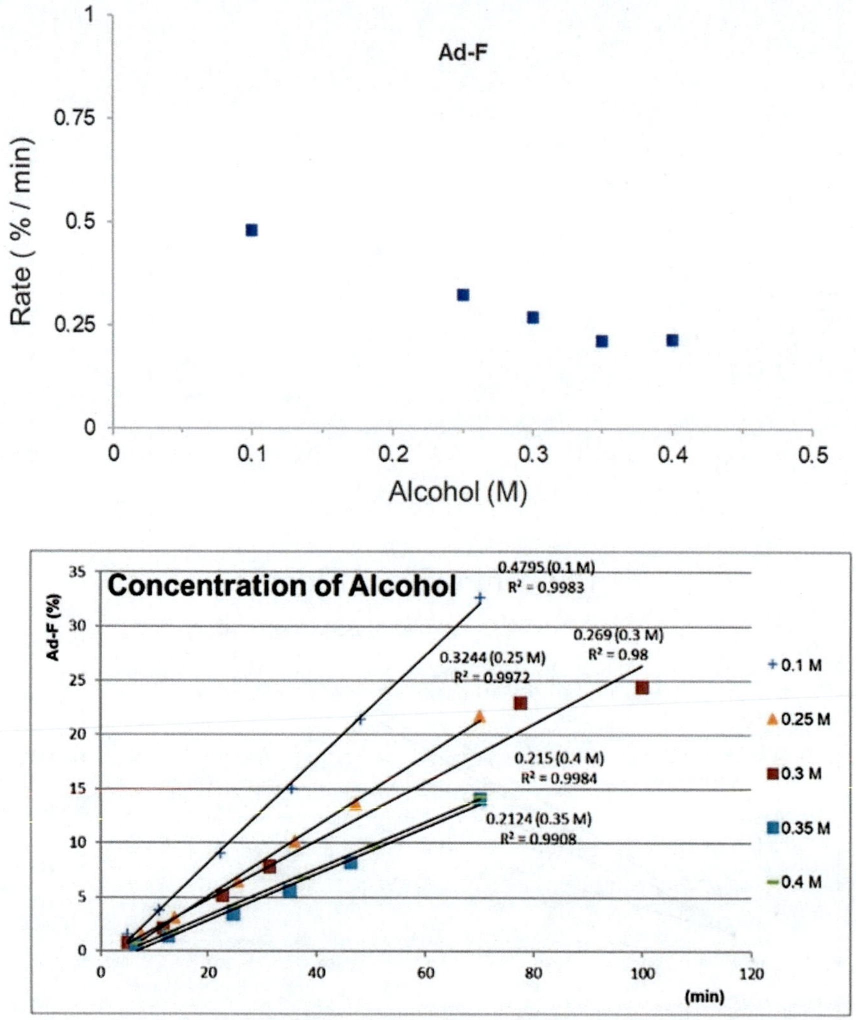

Fig. 4.26 (Left) Dependence of reaction rate, formation of Ad-F, on the concentration of alcohol. (Right) Plot of yield (Ad-F) in a given period of time under varying concentrations of alcohol. The yields of Ad-F were calculated based on the amount of S16b

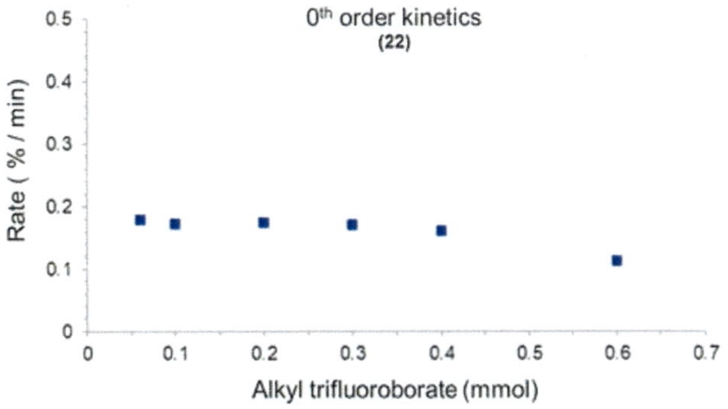

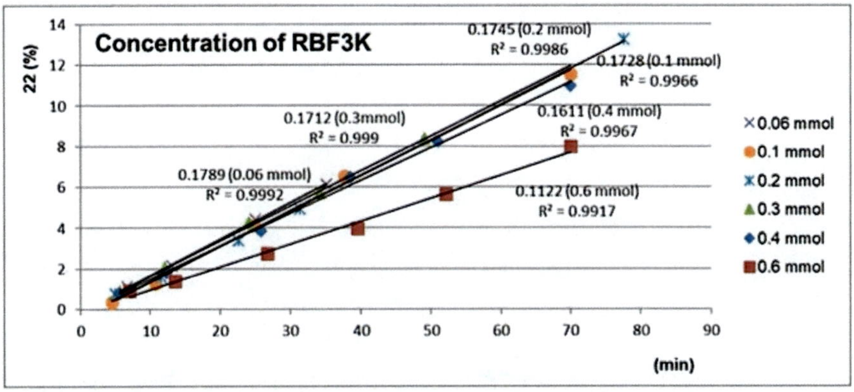

Fig. 4.27 (Left) Dependence of reaction rate, formation of 22, on the concentration of alkyl trifluoroborate. The result shows zero-order kinetics with respect to the alkyl trifluoroborate concentration. Due to the relatively insignificant formation of fluorination product (Ad-F), only the alcohol product was analyzed. (Right) Plot of yield (22) in a given period of time under varying concentrations of alkyl trifluoroborate S16b. The yield of 22 was calculated based on the amount of alcohol

4.4.11 Experimental Procedures and Characterization Data

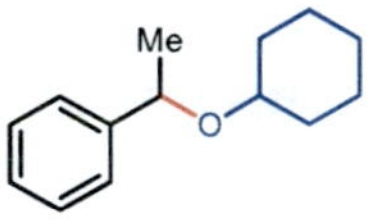

(1-(Cyclohexyloxy)ethyl)benzene (1)
General procedure was applied with alkyl trifluoroborate **S1b** (0.2 mmol), cyclohexanol (0.6 mmol), and $^{n}Bu_4NClO_4$ (0.2 mmol). Purification by flash column chromatography (silica gel, hexanes:CH_2Cl_2 = 7:3) afforded **1** (23.8 mg, 58%) as a

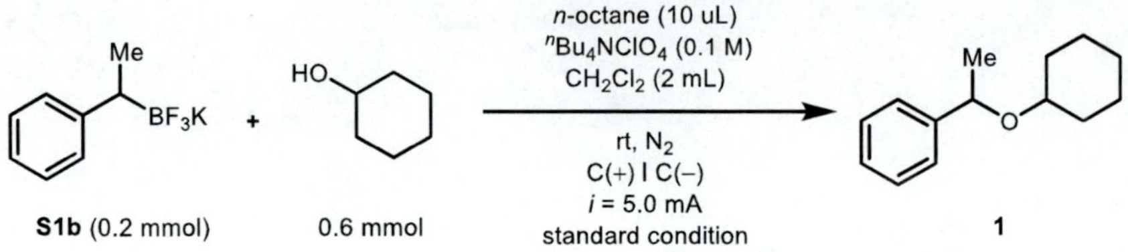

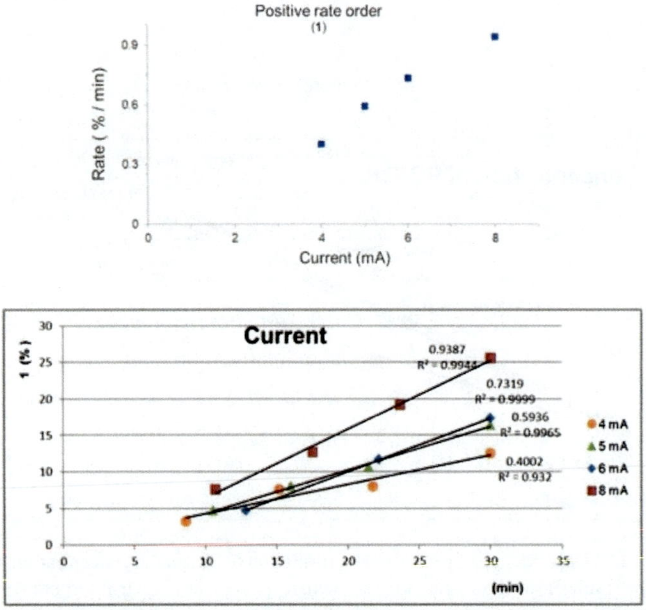

Fig. 4.28 (Left) Dependence of reaction rate, formation 1, on the current of the cell. The result shows near first-order kinetics with respect to the applied current. Due to the unchanged ratio of fluoride and alcohol nucleophile, only the alcohol product was analyzed. (Right) Plot of yield (1) in a given time period under varying currents. The yields of 1 were calculated based on the amount of S1b

colorless oil; $R_f = 0.4$ (hexanes:CH_2Cl_2 = 3:2). Spectroscopic data matches with previously reported data [36].

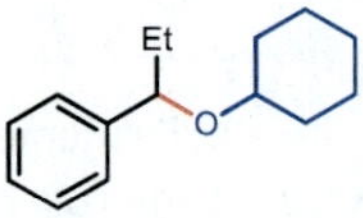

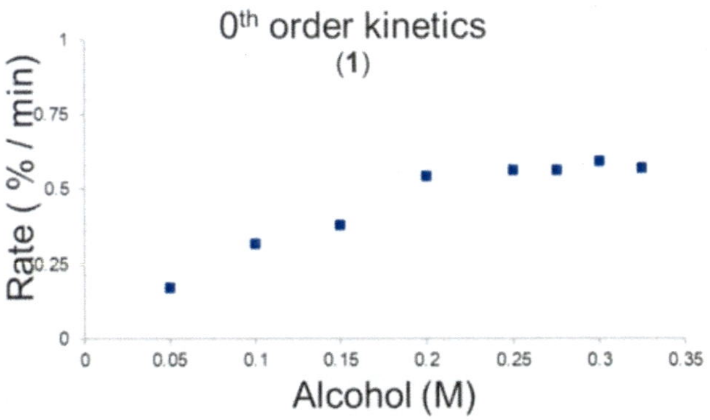

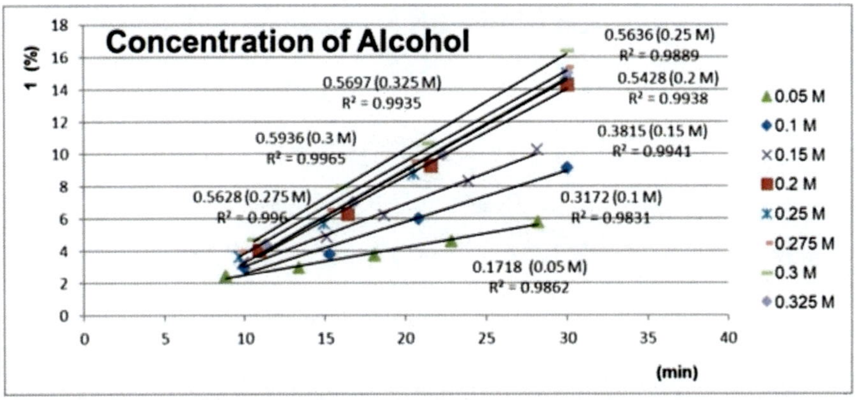

Fig. 4.29 (Left) Dependence of reaction rate, formation 1, on the concentration of alcohol. The result shows zero-order kinetics with respect to the alcohol concentration, especially near the synthetically relevant conditions (0.3 M). The formation of minor side products was disregarded. (Right) Plot of yield (1) in a given period of time under varying concentrations of alcohol. The yield of 1 was calculated based on the amount of S1b

(1-(Cyclohexyloxy)propyl)benzene (2)

General procedure was applied with alkyl trifluoroborate **S2b** (0.2 mmol), cyclohexanol (0.6 mmol), and ^{n}Bu$_4$NClO$_4$ (0.2 mmol). Purification by flash column chromatography (silica gel, hexanes:CH_2Cl_2 = 7:3) afforded **2** (23.6 mg, 54%) as a colorless oil; R_f = 0.44 (hexanes:EtOAc = 47:3); ^{1}H NMR (400 MHz, $CDCl_3$): δ 7.40–7.21 (m, 5H), 4.29 (dd, J = 7.7, 5.6 Hz, 1H), 3.21–3.11 (m, 1H), 2.02–1.91 (m, 1H), 1.83–1.57 (m, 5H), 1.55–1.42 (m, 1H), 1.39–1.25 (m, 2H), 1.25–1.09 (m, 3H), 0.92 (t, J = 7.4 Hz, 3H); ^{13}C NMR (126 MHz, $CDCl_3$): δ 144.3, 128.3, 127.2, 126.7, 80.5, 75.1, 33.8, 32.0, 31.7, 26.0, 24.5, 24.3, 10.8; HRMS (ESI) m/z: $[M+Na]^+$ calcd for $C_{15}H_{22}ONa$ 241.1568; found 241.1567.

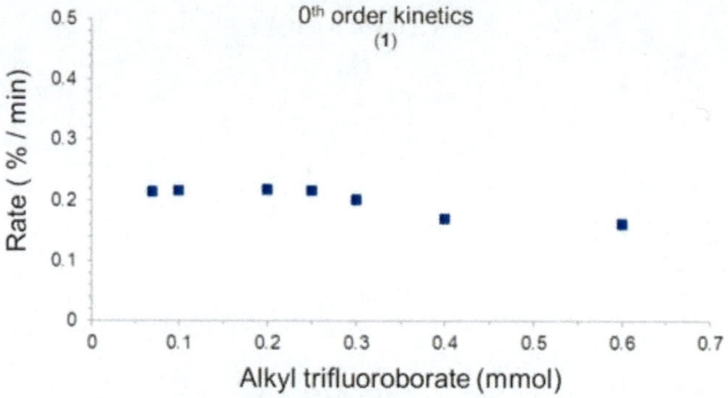

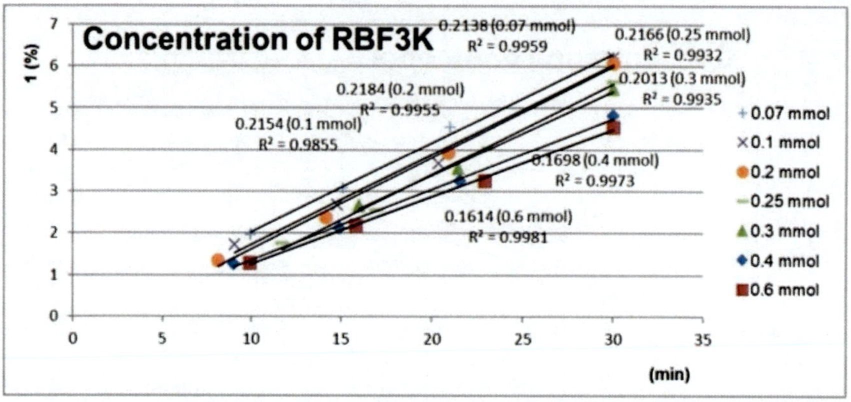

Fig. 4.30 (Left) Dependence of reaction rate, formation of 1, on the concentration of alkyl trifluoroborate. The result shows zero-order kinetics with respect to the alkyl trifluoroborate concentration. Due to the relatively insignificant formation of side products, only the alcohol product was analyzed. (Right) Plot of yield (1) in a given period of time under varying concentrations of alkyl trifluoroborate S1b. The yields of 1 were calculated based on the amount of alcohol

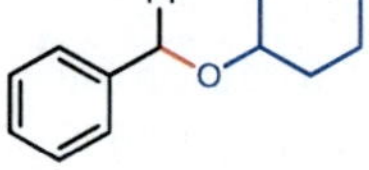

(1-(Cyclohexyloxy)-2-methylpropyl)benzene (3)

General procedure was applied with alkyl trifluoroborate **S3b** (0.2 mmol), cyclohexanol (0.6 mmol), and $^{n}Bu_4NClO_4$ (0.2 mmol). Purification by flash column chromatography (silica gel, hexanes:CH_2Cl_2 = 7:3) afforded **3** (22.0 mg, 47%) as a colorless oil; R_f = 0.56 (hexanes:EtOAc = 47:3); ^{1}H NMR (400 MHz, $CDCl_3$): δ 7.38–7.20 (m, 5H), 3.99 (d, J = 7.4 Hz, 1H), 3.17–3.05 (m, 1H), 1.98–1.77 (m, 2H), 1.77–1.58 (m, 3H), 1.52–1.40 (m, 1H), 1.38–1.23 (m, 2H), 1.23–1.08 (m, 3H), 1.00

(d, $J = 6.6$ Hz, 3H), 0.72 (d, $J = 6.7$ Hz, 3H); ^{13}C NMR (126 MHz, $CDCl_3$): δ 143.1, 128.0, 127.5, 127.1, 84.7, 74.9, 35.2, 33.8, 31.4, 26.0, 24.3, 24.1, 19.5, 19.3; HRMS (ESI) *m*/*z*: $[M+Na]^+$ calcd for $C_{16}H_{24}ONa$ 255.1725; found 255.1723.

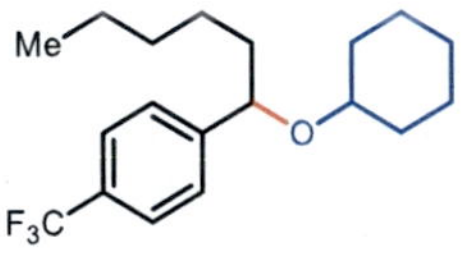

1-(1-(Cyclohexyloxy)hexyl)-4-(trifluoromethyl)benzene (4)

General procedure was applied with alkyl trifluoroborate **S25b** (0.2 mmol), cyclohexanol (0.6 mmol), and $^nBu_4NClO_4$ (0.2 mmol). Purification by flash column chromatography (silica gel, hexanes:CH_2Cl_2 = 6:1) afforded **4** (11.9 mg, 18%) as a colorless oil; $R_f = 0.6$ (hexanes:CH_2Cl_2 = 2:1); ^{1}H NMR (400 MHz, $CDCl_3$): δ 7.58 (d, $J = 8.0$ Hz, 2H), 7.42 (d, $J = 8.0$ Hz, 2H), 4.41 (dd, $J = 8.3$, 4.9 Hz, 1H), 3.14–3.07 (m, 1H), 2.00–1.86 (m, 1H), 1.85–1.62 (m, 4H), 1.62–1.40 (m, 3H), 1.40–1.07 (m, 10H), 0.95–0.79 (m, 3H); ^{13}C NMR (101 MHz, $CDCl_3$): δ 148.9, 129.4 (q, $J_{C\text{–}F} = 32.0$ Hz), 126.9, 125.3 (q, $J_{C\text{–}F} = 3.8$ Hz), 124.4 (q, $J_{C\text{–}F} = 271.5$ Hz), 78.6, 75.6, 39.1, 33.7, 31.8, 31.7, 25.9, 25.8, 24.4, 24.2, 22.7, 14.2; ^{19}F NMR (376 MHz, $CDCl_3$): δ −61.9; HRMS (ESI) *m*/*z*: $[M+H]^+$ calcd for $C_{19}H_{27}F_3ONa$ 351.1912; found 351.1910.

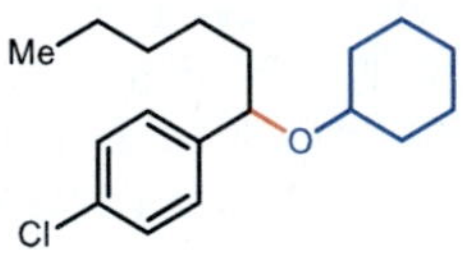

1-Chloro-4-(1-(cyclohexyloxy)hexyl)benzene (5)

General procedure was applied with alkyl trifluoroborate **S26b** (0.2 mmol), cyclohexanol (0.6 mmol), and $^nBu_4NClO_4$ (0.2 mmol). Purification by flash column chromatography (silica gel, hexanes:CH_2Cl_2 = 8:1) afforded **5** (24.8 mg, 42%) as a colorless oil; $R_f = 0.7$ (hexanes:CH_2Cl_2 = 2:1); ^{1}H NMR (400 MHz, $CDCl_3$): δ 7.29 (d, $J = 8.2$ Hz, 2H), 7.23 (d, $J = 8.5$ Hz, 2H), 4.32 (dd, $J = 8.1$, 5.2 Hz, 1H), 3.11–3.07 (m, 1H), 1.98–1.86 (m, 1H), 1.77–1.61 (m, 4H), 1.56–1.36 (m, 3H), 1.36–1.06 (m, 10H), 0.93–0.81 (m, 3H); ^{13}C NMR (101 MHz, $CDCl_3$): δ 143.0, 132.6, 128.3, 127.8, 78.2, 75.1, 38.9, 33.6, 31.7, 31.5, 25.8, 25.7, 24.3, 24.1, 22.6, 14.0; HRMS (ESI) *m*/*z*: $[M+Na]^+$ calcd for $C_{18}H_{27}ClONa$ 317.1648; found 317.1645.

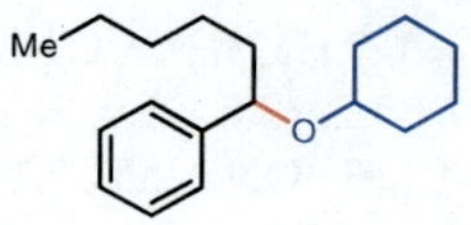

(1-(Cyclohexyloxy)hexyl)benzene (6)
General procedure was applied with alkyl trifluoroborate **S27b** (0.2 mmol), cyclohexanol (0.6 mmol), and $^{n}Bu_4NClO_4$ (0.2 mmol). Purification by flash column chromatography (silica gel, hexanes:CH_2Cl_2 = 6:1) afforded **6** (30.1 mg, 58%) as a colorless oil; R_f = 0.6 (hexanes:CH_2Cl_2 = 2:1); ^{1}H NMR (400 MHz, $CDCl_3$): δ 7.34–7.27 (m, 4H), 7.27–7.21 (m, 1H), 4.34 (dd, *J* = 8.2, 5.1 Hz, 1H), 3.14–3.09 (m, 1H), 1.99–1.88 (m, 1H), 1.79–1.61 (m, 4H), 1.57–1.38 (m, 3H), 1.37–1.07 (m, 10H), 0.90–0.80 (m, 3H); ^{13}C NMR (101 MHz, $CDCl_3$): δ 144.6, 128.3, 127.2, 126.7, 79.0, 75.0, 39.2, 33.8, 31.9, 31.7, 26.0, 26.0, 24.5, 24.3, 22.8, 14.2; HRMS (ESI) *m/z*: $[M+Na]^+$ calcd for $C_{18}H_{28}ONa$ 283.2038; found 283.2036.

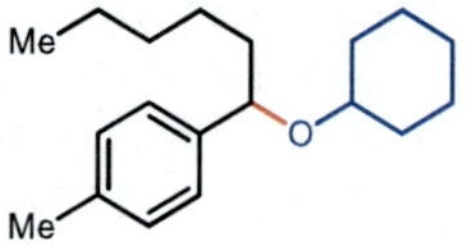

1-(1-(Cyclohexyloxy)hexyl)-4-methylbenzene (7)
General procedure was applied with alkyl trifluoroborate **S28b** (0.2 mmol), cyclohexanol (0.6 mmol), and $^{n}Bu_4NClO_4$ (0.2 mmol). Purification by flash column chromatography (silica gel, hexanes:CH_2Cl_2 = 2:1) afforded **7** (32.6 mg, 59%) as a colorless oil; R_f = 0.6 (hexanes:CH_2Cl_2 = 2:1); ^{1}H NMR (400 MHz, $CDCl_3$): δ 7.19 (d, *J* = 7.8 Hz, 2H), 7.13 (d, *J* = 7.8 Hz, 2H), 4.34–4.31 (m, 1H), 3.15–3.09 (m, 1H), 2.34 (s, 3H), 1.96–1.92 (m, 1H), 1.75–1.63 (m, 3H), 1.57–1.40 (m, 3H), 1.35–1.11 (m, 11H), 0.90–0.83 (m, 3H); ^{13}C NMR (101 MHz, $CDCl_3$): δ 141.3, 136.5, 128.8, 126.4, 78.6, 74.7, 39.1, 33.7, 31.8, 31.5, 25.8, 24.4, 24.1, 22.6, 21.10, 14.1; HRMS (ESI) *m/z*: $[M+Na]^+$ calcd for $C_{19}H_{30}ONa$ 297.2194; found 297.2196.

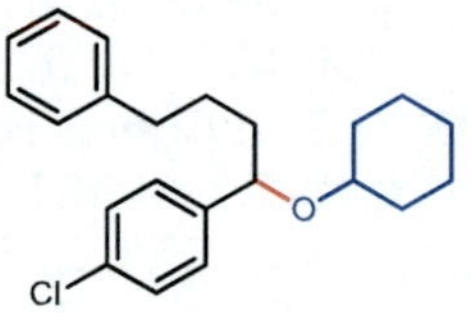

1-Chloro-4-(1-(cyclohexyloxy)-4-phenylbutyl)benzene (8)
General procedure was applied with alkyl trifluoroborate **S29b** (0.2 mmol), cyclohexanol (0.6 mmol), and $^{n}Bu_4NClO_4$ (0.2 mmol). Purification by flash column chromatography (silica gel, hexanes:EtOAc = 30:1) afforded **8** (20.7 mg, 30%) as a

colorless oil; $R_f = 0.7$ (hexanes:EtOAc = 10:1); ^{1}H NMR (400 MHz, $CDCl_3$): δ 7.34–7.08 (m, 9H), 4.33 (dd, $J = 8.0$, 4.7 Hz, 1H), 3.16–2.99 (m, 1H), 2.65–2.52 (m, 2H), 1.89 (d, $J = 12.3$ Hz, 1H), 1.84–1.41 (m, 7H), 1.34–1.03 (m, 6H); ^{13}C NMR (101 MHz, $CDCl_3$): δ 142.9, 142.6, 132.8, 128.6, 128.5, 128.4, 128.0, 125.8, 78.2, 75.3, 38.7, 36.0, 33.7, 31.7, 28.1, 25.9, 24.4, 24.2; HRMS (ESI) *m*/*z*: $[M+Na]^+$ calcd for $C_{22}H_{27}ClONa$ 365.1648; found 365.1644.

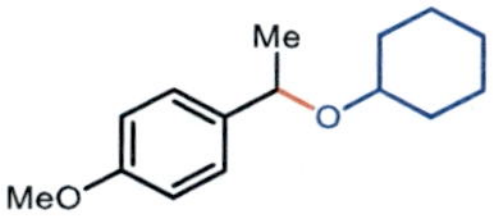

1-(1-(Cyclohexyloxy)ethyl)-4-methoxybenzene (9)

General procedure was applied with alkyl trifluoroborate **S6b** (0.2 mmol), cyclohexanol (0.6 mmol), and $^n Bu_4NClO_4$ (0.2 mmol). The reaction mixture was electrolyzed at a constant current of 5.0 mA for 2.0 h. Purification by flash column chromatography (silica gel, hexanes:CH_2Cl_2 = 7:3) afforded **9** (18.7 mg, 40%) as a colorless oil; $R_f = 0.46$ (hexanes:EtOAc = 10:1). Spectroscopic data matches with previously reported data [37].

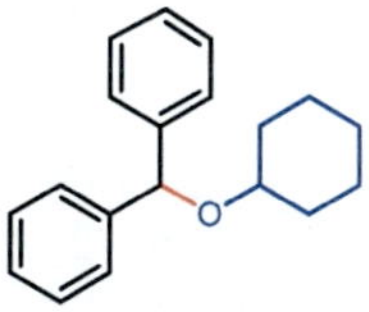

((Cyclohexyloxy)methylene)dibenzene (10)

General procedure was applied with alkyl trifluoroborate **S4b** (0.2 mmol), cyclohexanol (0.6 mmol), and $^n Bu_4NClO_4$ (0.2 mmol). Purification by flash column chromatography (silica gel, hexanes:CH_2Cl_2 = 6:1) afforded **10** (32.1 mg, 60%) as a colorless oil; $R_f = 0.6$ (hexanes:CH_2Cl_2 = 2:1); ^{1}H NMR (400 MHz, $CDCl_3$): δ 7.39–7.26 (m, 8H), 7.26–7.19 (m, 2H), 5.54 (s, 1H), 3.39–3.32 (m, 1H), 1.98–1.87 (m, 2H), 1.75–1.72 (m, 2H), 1.52–1.38 (m, 3H), 1.28–1.14 (m, 3H); ^{13}C NMR (101 MHz, $CDCl_3$): δ 143.3, 128.4, 127.3, 127.3, 80.1, 75.2, 32.5, 26.0, 24.3; HRMS (ESI) *m*/*z*: $[M+Na]^+$ calcd for $C_{19}H_{22}ONa$ 289.1568; found 289.1569.

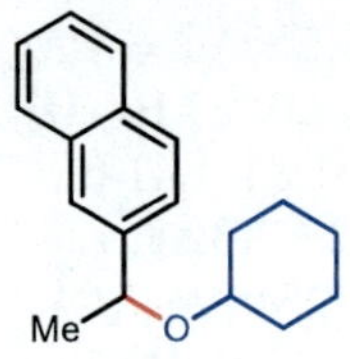

2-(1-(Cyclohexyloxy)ethyl)naphthalene (11)

General procedure was applied with alkyl trifluoroborate **S5b** (0.2 mmol), cyclohexanol (0.6 mmol), and $^{n}Bu_4NClO_4$ (0.2 mmol). Purification by flash column chromatography (silica gel, hexanes:EtOAc = 30:1) afforded **11** (29.2 mg, 57%) as a colorless oil; R_f = 0.4 (hexanes:EtOAc = 10:1); ^{1}H NMR (400 MHz, $CDCl_3$): δ 7.85–7.83 (m, 3H), 7.75–7.75 (m, 1H), 7.54–7.42 (m, 3H), 4.77 (q, J = 6.5 Hz, 1H), 3.24–3.17 (m, 1H), 2.09–1.98 (m, 1H), 1.82–1.62 (m, 3H), 1.50 (d, J = 6.5 Hz, 4H), 1.41–1.23 (m, 2H), 1.19–1.09 (m, 3H); ^{13}C NMR (101 MHz, $CDCl_3$): δ 142.7, 133.4, 133.1, 128.3, 128.0, 127.8, 126.1, 125.7, 124.9, 124.9, 124.6, 124.5, 75.1, 74.6, 74.5, 33.7, 32.0, 26.0, 25.0, 24.6, 24.4; HRMS (ESI) *m*/*z*: $[M+Na]^+$ calcd for $C_{18}H_{22}ONa$ 277.1568; found 277.1565.

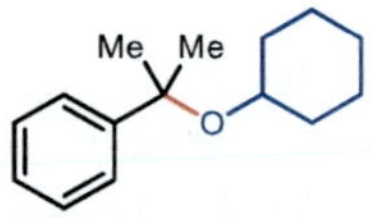

(2-(Cyclohexyloxy)propan-2-yl)benzene (12)

General procedure was applied with alkyl trifluoroborate **S9b** (0.2 mmol), cyclohexanol (0.6 mmol), and $^{n}Bu_4NClO_4$ (0.2 mmol). The reaction mixture was electrolyzed at a constant current of 5.0 mA for 1.5 h. Purification by flash column chromatography (silica gel, hexanes:CH_2Cl_2 = 3:1) afforded **12** (19.6 mg, 45%) as a colorless oil; R_f = 0.59 (hexanes:EtOAc = 9:1); ^{1}H NMR (500 MHz, $CDCl_3$): δ 7.50 (d, J = 7.3 Hz, 2H), 7.33 (t, J = 7.7 Hz, 2H), 7.24 (t, J = 7.3 Hz, 1H), 3.17–3.09 (m, 1H), 1.78–1.60 (m, 4H), 1.57–1.53 (m, 6H), 1.50–1.42 (m, J = 11.2, 6.5 Hz, 1H), 1.34–1.22 (m, 2H), 1.15–1.03 (m, 3H); ^{13}C NMR (126 MHz, $CDCl_3$): δ 147.4, 127.9, 126.9, 126.3, 76.7, 71.7, 35.3, 29.2, 25.8, 25.0; HRMS (ESI) *m*/*z*: $[M+Na]^+$ calcd for $C_{15}H_{22}ONa$ 241.1568; found 241.1565.

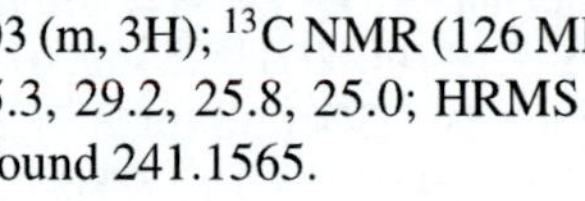

(4-((1-Phenylethoxy)methyl)phenyl)methanol (13)
General procedure was applied with alkyl trifluoroborate **S1b** (0.2 mmol), 1,4-phenylenedimethanol (0.6 mmol), and $^{n}Bu_4NClO_4$ (0.2 mmol). The reaction mixture was electrolyzed at a constant current of 5.0 mA for 2.0 h. Purification by flash column chromatography (silica gel, hexanes:EtOAc = 9:1) afforded **13** (22.4 mg, 46%) as a colorless oil; R_f = 0.4 (hexanes:EtOAc = 3:1); ^{1}H NMR (400 MHz, $CDCl_3$): δ 7.39–7.29 (m, 9H), 4.68 (s, 2H), 4.52–4.47 (q, *J* = 6.5 Hz, 1H), 4.47–4.44 (m, 1H), 4.32–4.29 (m, 1H), 1.67 (br, 1H), 1.48 (d, *J* = 6.5 Hz, 3H); ^{13}C NMR (101 MHz, $CDCl_3$): δ 143.8, 140.3, 138.3, 128.7, 128.1, 127.7, 127.22, 126.5, 77.3, 70.1, 65.3, 24.3; HRMS (ESI) *m*/*z*: $[M+Na]^+$ calcd for $C_{16}H_{18}O_2Na$ 265.1205; found 265.1203.

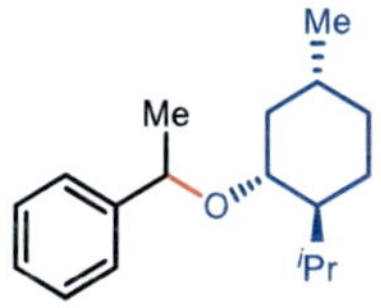

(1-(((1*R*,2*S*,5*R*)-2-Isopropyl-5-methylcyclohexyl)oxy)ethyl)benzene (14)
General procedure was applied with alkyl trifluoroborate **S1b** (0.2 mmol), (1*R*,2*S*,5*R*)-2-isopropyl-5-methylcyclohexan-1-ol (0.6 mmol), and $^{n}Bu_4NClO_4$ (0.2 mmol). The reaction mixture was electrolyzed at a constant current of 5.0 mA for 2.0 h. Purification by flash column chromatography (silica gel, hexanes:CH_2Cl_2 = 4:1) afforded **14** (27.0 mg, 52%) as a colorless oil; R_f = 0.42 (hexanes:CH_2Cl_2 = 4:1); ^{1}H NMR (500 MHz, $CDCl_3$, for both diastereomers) (the integration at 4.58 ppm and 4.51 ppm indicated the ratio of the two isomers of **14** to be 1:1): δ 7.42–7.20 (m, 5H), 4.58 (q, *J* = 6.5 Hz, 0.5H), 4.51 (q, *J* = 6.4 Hz, 0.5H), 3.20–3.15 (m, 0.5H), 2.94–2.89 (m, 0.5H), 2.36–2.33 (m, 0.5H), 2.30–2.23 (m, 0.5H), 2.21 (d, *J* = 11.9 Hz, 0.5H), 1.74 (d, *J* = 12.0 Hz, 0.5H), 1.68–1.51 (m, 2H), 1.45 (d, *J* = 6.5 Hz, 1.5H), 1.43 (d, *J* = 6.4 Hz, 1.5H), 1.34–1.16 (m, 2.5H), 0.95 (d, *J* = 7.1 Hz, 1.5H), 0.93 (d, *J* = 6.5 Hz, 1.5H), 0.90–0.78 (m, 2.5H), 0.86 (d, J = 7.9 Hz, 3.0H), 0.80 (d, J = 6.5 Hz, 1.5H), 0.28 (d, *J* = 6.9 Hz, 1.5H); ^{13}C NMR (101 MHz, $CDCl_3$, for both diastereomers): δ 145.6, 144.2, 128.3, 127.5, 127.3, 127.1, 126.4, 78.1, 76.9, 75.3, 74.0, 49.2, 48.6, 42.2, 40.5, 34.7, 34.6, 31.8, 31.6, 25.5, 24.9, 24.7, 23.7, 23.3, 22.9, 22.6, 22.4, 21.4, 16.3, 15.5; HRMS (ESI) *m*/*z*: $[M+Na]^+$ calcd for $C_{18}H_{28}ONa$ 283.2038; found 283.2034.

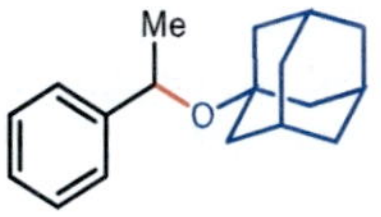

1-(1-Phenylethoxy)adamantane (15)

General procedure was applied with alkyl trifluoroborate **S1b** (0.2 mmol), adamantan-1-ol (0.6 mmol), and $^{n}Bu_4NClO_4$ (0.2 mmol). Purification by flash column chromatography (silica gel, hexanes:CH_2Cl_2 = 7:3) afforded **15** (29.3 mg, 57%) as a colorless oil; R_f = 0.52 (hexanes:EtOAc = 9:1). Spectroscopic data matches with previously reported data [38].

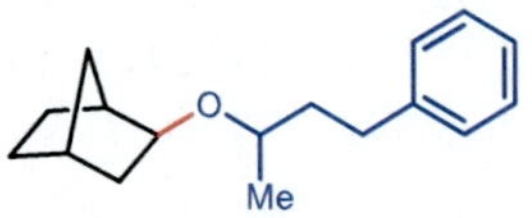

2-((4-Phenylbutan-2-yl)oxy)bicyclo[2.2.1]heptane (16)

General procedure was applied with alkyl trifluoroborate **S15b** (0.2 mmol), 4-phenylbutan-2-ol (0.6 mmol), and $^{n}Bu_4NPF_6$ (0.2 mmol). Purification by flash column chromatography (silica gel, hexanes:CH_2Cl_2 = 7:3) afforded **16** (37.3 mg, 76%) as a colorless oil; R_f = 0.57 (hexanes:EtOAc = 9:1); ^{1}H NMR (500 MHz, $CDCl_3$, for both diastereomers) (the integration at 1.19 ppm and 1.16 ppm indicated the ratio of the two isomers of **16** to be 1:1): δ 7.33–7.25 (m, 2H), 7.25–7.16 (m, 3H), 3.54–3.41 (m, 2H), 2.85–2.57 (m, 2H), 2.26 (s, 2H), 1.90–1.77 (m, 1H), 1.76–1.65 (m, 1H), 1.65–1.35 (m, 5H), 1.19 (d, J = 6.1 Hz, 1.5H), 1.16 (d, J = 6.1 Hz, 1.5H), 1.14–0.94 (m, 3H); ^{13}C NMR (126 MHz, $CDCl_3$, for both diastereomers): δ 142.8, 142.7, 128.5, 128.5, 128.4, 125.7, 125.7, 80.2, 79.7, 72.3, 71.9, 42.0, 40.7, 40.5, 40.1, 39.1, 39.0, 35.4, 35.2, 35.1, 35.0, 32.3, 32.2, 28.8, 24.9, 24.8, 20.7, 20.4; HRMS (ESI) *m*/*z*: $[M+Na]^+$ calcd for $C_{17}H_{24}ONa$ 267.1725; found 267.1722.

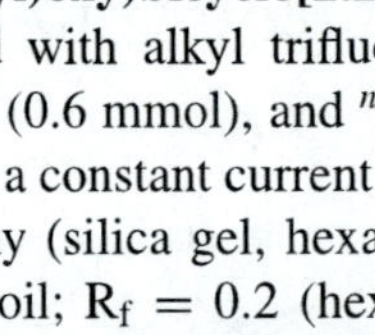

2-((1-(Phenylethynyl)cyclohexyl)oxy)bicyclo[2.2.1]heptane (17)

General procedure was applied with alkyl trifluoroborate **S15b** (0.2 mmol), 1-(phenylethynyl)cyclohexan-1-ol (0.6 mmol), and $^{n}Bu_4NPF_6$ (0.2 mmol). The reaction mixture was electrolyzed at a constant current of 5.0 mA for 2.0 h. Purification by flash column chromatography (silica gel, hexanes:CH_2Cl_2 = 5:1) afforded **17** (18.7 mg, 32%) as a colorless oil; R_f = 0.2 (hexanes:CH_2Cl_2 = 5:1); ^{1}H NMR (400 MHz, $CDCl_3$): δ 7.44–7.42 (m, 2H), 7.31–7.30 (m, 3H), 3.88 (d, *J* = 7.2 Hz, 1H), 2.32 (d, *J* = 4.5 Hz, 1H), 2.21 (s, 1H), 2.03–1.93 (m, 2H), 1.73–1.38 (m, 11H),

1.29–1.26 (m, 2H), 1.14–0.99 (m, 3H); ^{13}C NMR (101 MHz, $CDCl_3$): δ 131.7, 128.4, 128.1, 123.6, 92.4, 85.6, 77.1, 74.1, 43.4, 42.5, 38.7, 38.6, 35.4, 35.4, 28.5, 25.7, 25.1, 23.3, 23.3; HRMS (ESI) *m*/*z*: $[M+Na]^+$ calcd for $C_{21}H_{26}ONa$ 317.1881; found 317.1879.

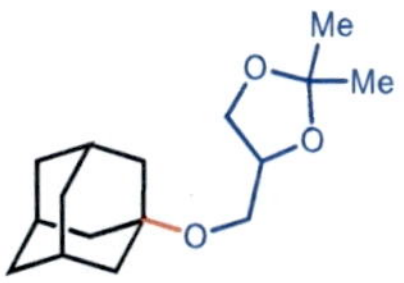

4-(((Adamantan-1-yl)oxy)methyl)-2,2-dimethyl-1,3-dioxolane (18)
General procedure was applied with alkyl trifluoroborate **S16b** (0.2 mmol), (2,2-dimethyl-1,3-dioxolan-4-yl)methanol (1.2 mmol), and $^{n}Bu_4NPF_6$ (0.2 mmol). Purification by flash column chromatography (silica gel, hexanes:EtOAc = 9:1) afforded **18** (24.5 mg, 46%) as a colorless oil; R_f = 0.37 (hexanes:EtOAc = 9:1); ^{1}H NMR (500 MHz, $CDCl_3$): δ 4.22–4.13 (m, 1H), 4.10–4.01 (m, 1H), 3.74 (dd, *J* = 8.1, 6.0 Hz, 1H), 3.52 (dd, *J* = 9.1, 5.3 Hz, 1H), 3.36 (dd, *J* = 8.9, 7.0 Hz, 1H), 2.18–2.08 (m, 3H), 1.77–1.69 (m, 6H), 1.67–1.53 (m, 6H), 1.41 (s, 3H), 1.35 (s, 3H); ^{13}C NMR (126 MHz, $CDCl_3$): δ 109.3, 75.4, 72.5, 67.6, 61.6, 41.6, 36.6, 30.6, 27.0, 25.5; HRMS (ESI) *m*/*z*: $[M+Na]^+$ calcd for $C_{16}H_{26}O_3Na$ 289.1780; found 289.1777.

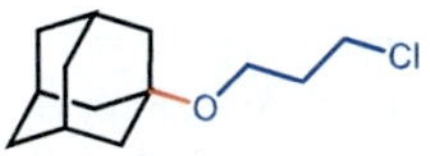

1-(3-Chloropropoxy)adamantane (19)
General procedure was applied with alkyl trifluoroborate **S16b** (0.2 mmol), 3-chloropropan-1-ol (1.2 mmol), and $^{n}Bu_4NPF_6$ (0.2 mmol). Purification by flash column chromatography (silica gel, hexanes:CH_2Cl_2 = 4:1) afforded **19** (21.2 mg, 46%) as a colorless oil; R_f = 0.7 (hexanes:EtOAc = 9:1); ^{1}H NMR (400 MHz, $CDCl_3$): δ 3.64 (t, *J* = 6.2 Hz, 2H), 3.54 (t, *J* = 5.8 Hz, 2H), 2.19–2.09 (m, 3H), 2.00–1.91 (m, 2H), 1.78–1.70 (m, 6H), 1.69–1.54 (m, 6H); ^{13}C NMR (101 MHz, $CDCl_3$): δ 72.2, 56.2, 42.5, 41.7, 36.6, 33.6, 30.6; HRMS (ESI) *m*/*z*: $[M+Na]^+$ calcd for $C_{13}H_{21}ClONa$ 251.1179; found 251.1176.

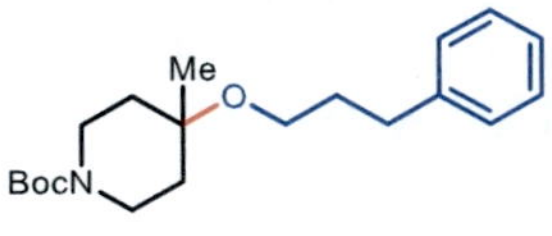

***tert*-butyl 4-methyl-4-(3-phenylpropoxy)piperidine-1-carboxylate (20)**
General procedure was applied with alkyl trifluoroborate **S18b** (0.2 mmol), 3-phenyl-1-propanol (0.6 mmol), and $^{n}Bu_4NPF_6$ (0.2 mmol). The reaction mixture was electrolyzed at a constant current of 5.0 mA for 3.0 h. Purification by flash column chromatography (silica gel, hexanes:CH_2Cl_2 = 4:1) afforded **20** (21.5 mg, 32%) as a colorless oil; R_f = 0.3 (hexanes:CH_2Cl_2 = 2:1); ^{1}H NMR (400 MHz, $CDCl_3$) δ 7.29–7.24 (m, 2H), 7.20–7.12 (m, 3H), 3.70–3.67 (m, 2H), 3.30 (t, J = 6.3 Hz, 2H), 3.18–3.05 (m, 2H), 2.73–2.64 (m, 2H), 1.91–1.81 (m, 2H), 1.75–1.66 (m, 2H), 1.45–1.43 (m, 11H), 1.12 (s, 3H); ^{13}C NMR (101 MHz, $CDCl_3$) δ 155.1, 142.3, 128.6, 128.4, 125.9, 79.4, 71.3, 59.9, 40.0, 35.8, 32.8, 32.2, 28.6, 24.7; HRMS (ESI) *m*/*z*: $[M+Na]^+$ calcd for $C_{20}H_{31}NO_3Na$ 356.2202; found 356.2201.

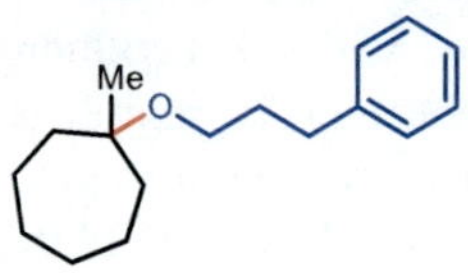

1-Methyl-1-(3-phenylpropoxy)cycloheptane (21)
General procedure was applied with alkyl trifluoroborate **S20b** (0.2 mmol), 3-phenyl-1-propanol (0.6 mmol), and $^{n}Bu_4NPF_6$ (0.2 mmol). The reaction mixture was electrolyzed at a constant current of 5.0 mA for 3.0 h. Purification by flash column chromatography (silica gel, hexanes:CH_2Cl_2 = 8:1) afforded **21** (19.7 mg, 40%) as a colorless oil; R_f = 0.5 (hexanes:CH_2Cl_2 = 3:1); ^{1}H NMR (400 MHz, $CDCl_3$) δ 7.29–7.25 (m, 2H), 7.23–7.14 (m, 3H), 3.33 (t, J = 6.4 Hz, 2H), 2.74–2.66 (m, 2H), 1.90–1.75 (m, 4H), 1.63–1.46 (m, 8H), 1.40–1.32 (m, 2H), 1.12 (s, 3H); ^{13}C NMR (101 MHz, $CDCl_3$) δ 142.4, 128.4, 128.2, 125.6, 77.4, 59.9, 39.6, 32.6, 32.2, 29.8, 26.3, 22.4; HRMS (ESI) *m*/*z*: $[M+Na]^+$ calcd for $C_{17}H_{26}ONa$ 269.1881; found 269.1879.

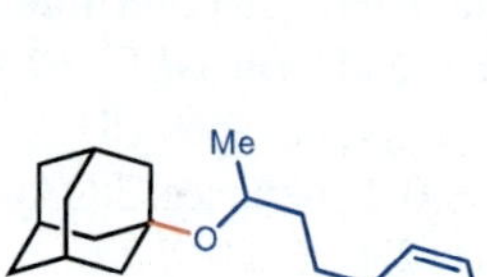

1-((4-Phenylbutan-2-yl)oxy)adamantane (22)
General procedure was applied with alkyl trifluoroborate **S16b** (0.2 mmol), 4-phenylbutan-2-ol (0.6 mmol), and $^{n}Bu_4NPF_6$ (0.2 mmol). Purification by flash column chromatography (silica gel, hexanes:CH_2Cl_2 = 7:3) afforded **22** (36.3 mg, 64%) as a colorless oil; R_f = 0.32 (hexanes:CH_2Cl_2 = 7:3); ^{1}H NMR (500 MHz, $CDCl_3$): δ 7.31–7.14 (m, 5H), 3.85–3.77 (m, 1H), 2.78–2.54 (m, 2H), 2.18–2.10 (m, 3H), 1.83–1.56 (m, 14H), 1.17 (d, J = 6.1 Hz, 3H); ^{13}C NMR (126 MHz, $CDCl_3$):

δ 142.9, 128.5, 128.4, 125.7, 72.6, 65.2, 43.0, 40.8, 36.7, 32.7, 30.8, 23.7; HRMS (ESI) *m/z*: $[M+Na]^+$ calcd for $C_{20}H_{28}ONa$ 307.2038; found 307.2048.

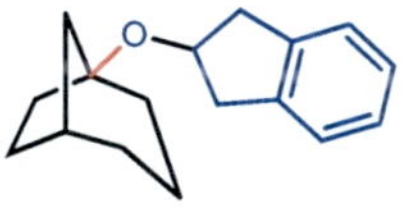

2-(Bicyclo[3.2.1]octan-1-yloxy)-2,3-dihydro-1*H*-indene (23)

General procedure was applied with alkyl trifluoroborate **S11b** (0.2 mmol), 2,3-dihydro-1*H*-inden-2-ol (0.6 mmol), and $^nBu_4NPF_6$ (0.2 mmol). Purification by flash column chromatography (silica gel, hexanes:EtOAc = 30:1) afforded **23** (13.2 mg, 27%) as a colorless oil; R_f = 0.5 (hexanes:EtOAc = 10:1); ^{1}H NMR (400 MHz, $CDCl_3$): δ 7.17–7.11 (m, 4H), 4.49 (p, *J* = 7.0 Hz, 1H), 3.15–3.09 (m, 2H), 2.94–2.88 (m, 2H), 2.23 (d, *J* = 5.6 Hz, 1H), 1.87–1.59 (m, 7H), 1.46–1.23 (m, 5H); ^{13}C NMR (101 MHz, $CDCl_3$): δ 141.2, 126.5, 124.6, 84.2, 74.9, 43.9, 41.8, 41.8, 37.6, 35.1, 33.9, 31.9, 27.7, 20.4; HRMS (ESI) *m/z*: $[M+Na]^+$ calcd for $C_{17}H_{22}ONa$ 265.1568; found 265.1566.

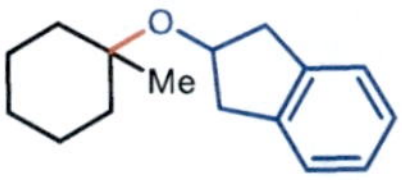

2-((1-Methylcyclohexyl)oxy)-2,3-dihydro-1*H*-indene (24)

General procedure was applied with alkyl trifluoroborate **S7b** (0.2 mmol), 2,3-dihydro-1*H*-inden-2-ol (0.6 mmol), and $^nBu_4NPF_6$ (0.2 mmol). Purification by flash column chromatography (silica gel, hexanes:EtOAc = 30:1) afforded **24** (19.0 mg, 41%) as a colorless oil; R_f = 0.7 (hexanes:EtOAc = 10:1); ^{1}H NMR (400 MHz, $CDCl_3$): δ 7.19–7.08 (m, 4H), 4.50 (p, *J* = 7.3 Hz, 1H), 3.14–3.08 (m, 2H), 2.95–2.90 (m, 2H), 1.76–1.58 (m, 5H), 1.45–1.27 (m, 5H), 1.19 (s, 3H); ^{13}C NMR (101 MHz, $CDCl_3$): δ 141.3, 126.5, 124.5, 74.5, 72.6, 41.9, 37.3, 26.2, 26.1, 22.6; HRMS (ESI) *m/z*: $[M+Na]^+$ calcd for $C_{16}H_{22}ONa$ 253.1568; found 253.1569.

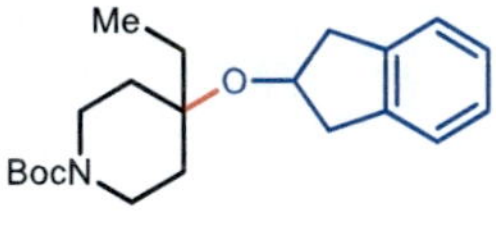

***tert*-butyl 4-((2,3-dihydro-1H-inden-2-yl)oxy)-4-ethylpiperidine-1-carboxylate (25)**

General procedure was applied with alkyl trifluoroborate **S19b** (0.2 mmol), 2,3-dihydro-1*H*-inden-2-ol (0.6 mmol), and $^nBu_4NPF_6$ (0.2 mmol). The reaction mixture was electrolyzed at a constant current of 5.0 mA for 3.0 h. Purification by flash column chromatography (silica gel, hexanes:CH_2Cl_2 = 4:1) afforded **25** (24.3 mg, 35%) as a colorless oil; R_f = 0.3 (hexanes:CH_2Cl_2 = 2:1); ^{1}H NMR (400 MHz, $CDCl_3$) δ 7.14 (s, 4H), 4.46 (p, J = 7.4 Hz, 1H), 3.77–3.74 (m, 2H), 3.25–3.04 (m, 4H), 2.99–2.93 (m, 2H), 1.74–1.70 (m, 2H), 1.56 (q, J = 7.4 Hz, 2H), 1.47–1.45 (m, 11H), 0.91 (t, J = 7.4 Hz, 3H); ^{13}C NMR (126 MHz, $CDCl_3$) δ 155.1, 140.9, 126.7, 124.5, 79.4, 74.9, 72.7, 41.4, 40.0, 34.0, 30.5, 28.6, 7.4; HRMS (ESI) m/z: $[M+Na]^+$ calcd for $C_{21}H_{31}NO_3Na$ 368.2202; found 368.2198.

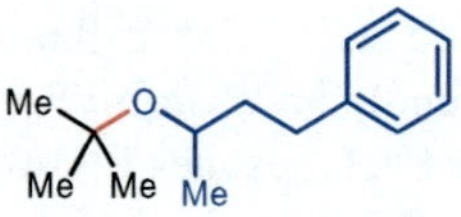

(3-(*tert*-butoxy)butyl)benzene (26)

General procedure was applied with alkyl trifluoroborate **S10b** (0.2 mmol), 4-phenylbutan-2-ol (0.6 mmol), and ^{n}Bu$_4$NPF$_6$ (0.2 mmol). Purification by flash column chromatography (silica gel, hexanes:EtOAc = 47:3) afforded **26** (24.2 mg, 59%) as a colorless oil; R_f = 0.52 (hexanes:EtOAc = 9:1). Spectroscopic data matches with previously reported data [39].

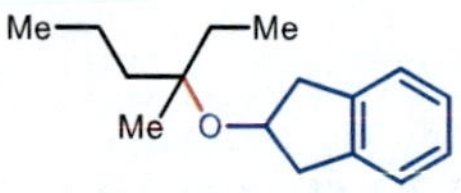

2-((3-Methylhexan-3-yl)oxy)-2,3-dihydro-1*H*-indene (27)

General procedure was applied with alkyl trifluoroborate **S21b** (0.2 mmol), 2,3-dihydro-1*H*-inden-2-ol (0.6 mmol), and ^{n}Bu$_4$NPF$_6$ (0.2 mmol). The reaction mixture was electrolyzed at a constant current of 5.0 mA for 3.0 h. Purification by flash column chromatography (silica gel, hexanes:CH_2Cl_2 = 10:1) afforded **27** (19.1 mg, 41%) as a colorless oil; R_f = 0.5 (hexanes:CH_2Cl_2 = 3:1); ^{1}H NMR (400 MHz, $CDCl_3$) δ 7.15–7.09 (m, 4H), 4.45 (p, J = 7.4 Hz, 1H), 3.10–3.04 (m, 2H), 2.91–2.85 (m, 2H), 1.53–1.31 (m, 6H), 1.12 (s, 3H), 0.96–0.80 (m, 6H); ^{13}C NMR (101 MHz, $CDCl_3$) δ 141.3, 126.5, 124.5, 77.9, 72.7, 41.7, 41.0, 31.3, 23.8, 17.1, 14.9, 8.4; HRMS (ESI) m/z: $[M+Na]^+$ calcd for $C_{16}H_{24}ONa$ 255.1725; found 255.1722.

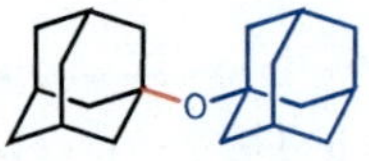

1,1'-Oxybis(adamantane) (28)

General procedure was applied with alkyl trifluoroborate **S16b** (0.2 mmol), adamantan-1-ol (0.6 mmol), and $^{n}Bu_4NPF_6$ (0.2 mmol). Purification by flash column chromatography (silica gel, hexanes:EtOAc = 40:1) afforded **28** (25.3 mg, 44%) as a white solid; m.p. 158–160 °C; $R_f = 0.8$ (hexanes:EtOAc = 10:1); ^{1}H NMR (400 MHz, $CDCl_3$): δ 2.12–2.04 (m, 6H), 1.95–1.83 (m, 12H), 1.65–1.54 (m, 12H); ^{13}C NMR (101 MHz, $CDCl_3$): δ 74.4, 46.0, 36.6, 31.2; HRMS (ESI) *m*/*z*: $[M+Na]^+$ calcd for $C_{20}H_{30}ONa$ 309.2194; found 303.2192.

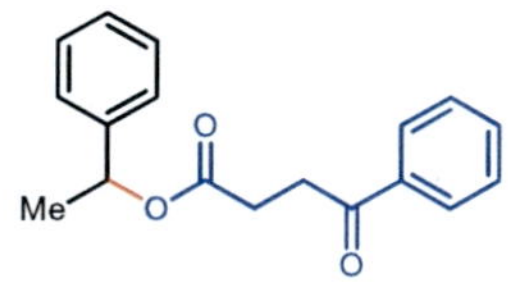

1-Phenylethyl 4-oxo-4-phenylbutanoate (29)

General procedure was applied with alkyl trifluoroborate **S1b** (0.2 mmol), 4-oxo-4-phenylbutanoic acid (0.6 mmol), and $^{n}Bu_4ClO_4$ (0.2 mmol). The reaction mixture was electrolyzed at a constant current of 5.0 mA for 2.0 h. Purification by flash column chromatography (silica gel, hexanes:EtOAc = 9:1) afforded **29** (27.3 mg, 48%) as a colorless oil; $R_f = 0.6$ (hexanes:EtOAc = 3:1); ^{1}H NMR (400 MHz, $CDCl_3$): δ 8.00–7.95 (m, 2H), 7.57 (t, $J = 7.4$ Hz, 1H), 7.46 (t, $J = 7.5$ Hz, 2H), 7.38–7.31 (m, 4H), 7.31–7.27 (m, 1H), 5.92 (q, $J = 6.6$ Hz, 1H), 3.43–3.20 (m, 2H), 2.91–2.69 (m, 2H), 1.55 (d, $J = 6.6$ Hz, 3H); ^{13}C NMR (101 MHz, $CDCl_3$): δ 198.2, 172.3, 141.8, 136.7, 133.3, 128.7, 128.6, 128.2, 128.0, 126.2, 72.8, 33.5, 28.7, 22.3; HRMS (ESI) *m*/*z*: $[M+Na]^+$ calcd for $C_{18}H_{18}O_3Na$ 305.1154; found 305.1170.

Adamantan-1-yl decanoate (30)

General procedure was applied with alkyl trifluoroborate **S16b** (0.2 mmol), decanoic acid (0.6 mmol), and $^{n}Bu_4NPF_6$ (0.2 mmol). The reaction mixture was electrolyzed at a constant current of 5.0 mA for 2.0 h. Purification by flash column chromatography (silica gel, hexanes:EtOAc = 50:1) afforded **30** (41.1 mg, 67%) as a colorless oil; R_f = 0.8 (hexanes:EtOAc = 10:1); ^{1}H NMR (400 MHz, $CDCl_3$): δ 2.18 (t, $J = 7.4$ Hz, 2H), 2.16–2.05 (m, 9H), 1.71–1.61 (m, 6H), 1.60–1.50 (m, 2H), 1.35–1.18 (m, 12H), 0.87 (t, $J = 6.7$ Hz, 3H); ^{13}C NMR (101 MHz, $CDCl_3$): δ 173.2, 80.1, 41.5, 36.4, 35.9, 32.0, 30.9, 29.6, 29.4, 29.4, 29.2, 25.3, 22.8, 14.2; HRMS (ESI) *m*/*z*: $[M+Na]^+$ calcd for $C_{20}H_{34}O_2Na$, 329.2457; found 329.2458.

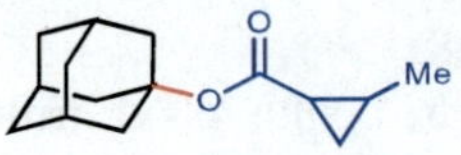

Adamantan-1-yl 2-methylcyclopropane-1-carboxylate (31)
General procedure was applied with alkyl trifluoroborate **S16b** (0.2 mmol), 2-methylcyclopropane-1-carboxylic acid (cis:trans = 1:1, 0.6 mmol), and $^{n}Bu_4NPF_6$ (0.2 mmol). The reaction mixture was electrolyzed at a constant current of 5.0 mA for 2.0 h. Purification by flash column chromatography (silica gel, hexanes:EtOAc = 40:1) afforded **31** (37.1 mg, 79%) as a colorless oil; $R_f = 0.6$ (hexanes:EtOAc = 10:1); ^{1}H NMR (400 MHz, for both diastereomers): δ 2.20–2.03 (m, 9H), 1.73–1.60 (m, 6H), 1.33–1.14 (m, 3H), 1.08 (d, $J = 6.1$ Hz, 3H), 0.59–0.56 (m, 1H); ^{13}C NMR (101 MHz, $CDCl_3$, for both diastereomers): δ 173.6, 80.0, 41.4, 36.2, 30.8, 22.4, 17.9, 16.7, 16.4; HRMS (ESI) *m*/*z*: $[M+H]^+$ calcd for $C_{15}H_{23}O_2$ 235.1698; found 235.1693.

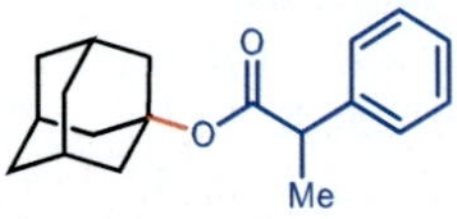

Adamantan-1-yl 2-phenylpropanoate (32)
General procedure was applied with alkyl trifluoroborate **S16b** (0.2 mmol), 2-phenylpropanoic acid (0.6 mmol), and $^{n}Bu_4NPF_6$ (0.2 mmol). The reaction mixture was electrolyzed at a constant current of 5.0 mA for 2.0 h. Purification by flash column chromatography (silica gel, hexanes:EtOAc = 50:1) afforded **32** (30.1 mg, 53%) as a colorless oil; $R_f = 0.8$ (hexanes:EtOAc = 10:1); ^{1}H NMR (400 MHz): δ 7.31–7.28 (m, 4H), 7.25–7.22 (m, 1H), 3.60 (q, $J = 7.1$ Hz, 1H), 2.17–2.10 (m, 3H), 2.09–2.01 (m, 6H), 1.68–1.60 (m, 6H), 1.44 (d, $J = 7.2$ Hz, 3H); ^{13}C NMR (101 MHz, $CDCl_3$): δ 173.8, 141.5, 128.6, 127.6, 126.9, 80.7, 46.7, 41.3, 36.3, 30.9, 18.7; HRMS (ESI) *m*/*z*: $[M+Na]^+$ calcd for $C_{19}H_{24}O_2Na$ 307.1674; found 307.1672.

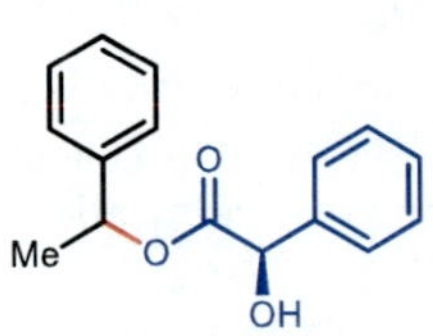

1-Phenylethyl (2*R*)-2-hydroxy-2-phenylacetate (33)
General procedure was applied with alkyl trifluoroborate **S1b** (0.2 mmol), (*R*)-2-hydroxy-2-phenylacetic acid (0.6 mmol), and $^{n}Bu_4ClO_4$ (0.2 mmol). The reaction mixture was electrolyzed at a constant current of 5.0 mA for 2.0 h. Purification by flash column chromatography (silica gel, hexanes:EtOAc = 10:1) afforded

33 (33.8 mg, 66%) as a white gum; R_f = 0.6 (hexanes:EtOAc = 3:1); ^{1}H NMR (500 MHz, $CDCl_3$, for both diastereomers) (the integration at 1.55 ppm and 1.41 ppm indicated the ratio of the two isomers of **29** to be 1:1): δ 7.49–7.44 (m, 1H), 7.42–7.29 (m, 6H), 7.24–7.13 (m, 2H), 6.97–6.93 (m, 1H), 5.92 (p, J = 6.5 Hz, 1H), 5.22 (s, 0.5H), 5.17 (s, 0.5H), 3.38 (br, 1H), 1.55 (d, J = 6.6 Hz, 1.5H), 1.41 (d, J = 6.6 Hz, 1.5H); ^{13}C NMR (126 MHz, $CDCl_3$, for both diastereomers): δ 173.2, 173.1, 140.9, 140.7, 138.5, 138.3, 128.8, 128.7, 128.6, 128.5, 128.5, 128.4, 128.0, 126.9, 126.6, 126.3, 125.5, 74.7, 74.6, 73.2, 73.1, 22.5, 21.8; HRMS (ESI) *m/z*: $[M+Na]^+$ calcd for $C_{16}H_{16}O_3Na$ 279.0997; found 279.0994.

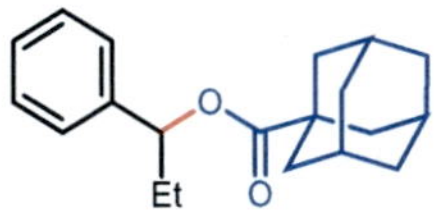

1-Phenylpropyl adamantane-1-carboxylate (34)

General procedure was applied with alkyl trifluoroborate **S2b** (0.2 mmol), adamantane-1-carboxylic acid (0.6 mmol), and $^nBu_4ClO_4$ (0.2 mmol). The reaction mixture was electrolyzed at a constant current of 5.0 mA for 2.0 h. Purification by flash column chromatography (silica gel, hexanes:EtOAc = 30:1) afforded **34** (22.7 mg, 38%) as a colorless oil; R_f = 0.5 (hexanes:EtOAc = 10:1); ^{1}H NMR (400 MHz, $CDCl_3$): δ 7.37–7.23 (m, 5H), 5.71–5.61 (m, 1H), 2.07–1.98 (m, 3H), 1.95–1.90 (m, 6H), 1.90–1.77 (m, 2H), 1.72–1.72 (m, 6H), 0.89 (t, J = 7.4 Hz, 3H); ^{13}C NMR (101 MHz, $CDCl_3$): δ 177.0, 141.3, 128.4, 127.7, 126.3, 76.6, 40.9, 39.0, 36.7, 29.9, 28.1, 10.0; HRMS (ESI) *m/z*: $[M+Na]^+$ calcd for $C_{20}H_{26}O_2Na$ 321.1831; found 321.1828.

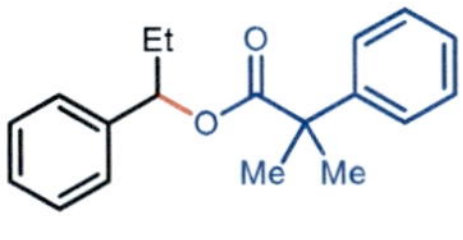

1-Phenylpropyl 2-methyl-2-phenylpropanoate (35)

General procedure was applied with alkyl trifluoroborate **S2b** (0.2 mmol), 2-methyl-2-phenylpropanoic acid (0.6 mmol), and $^nBu_4ClO_4$ (0.2 mmol). The reaction mixture was electrolyzed at a constant current of 5.0 mA for 2.0 h. Purification by flash column chromatography (silica gel, hexanes:EtOAc = 30:1) afforded **35** (24.5 mg, 43%) as a colorless oil; R_f = 0.5 (hexanes:EtOAc = 10:1); ^{1}H NMR (400 MHz, $CDCl_3$): δ 7.33–7.18 (m, 8H), 7.17–7.10 (m, 2H), 5.67–5.60 (m, 1H), 1.83–1.65 (m, 2H), 1.60–1.57 (m, 6H), 0.76 (t, J = 7.4 Hz, 3H); ^{13}C NMR (101 MHz, $CDCl_3$): δ 176.0, 144.7, 140.8, 128.4, 128.3, 127.7, 126.7, 126.3, 125.9, 77.7, 46.7, 29.7, 26.5, 26.4, 9.9; HRMS (ESI) *m/z*: $[M+Na]^+$ calcd for $C_{19}H_{22}O_2Na$ 305.1518; found 305.1514.

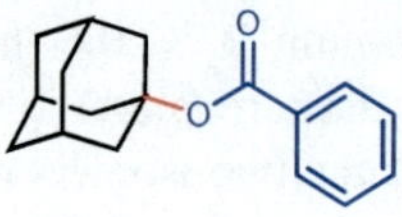

Adamantan-1-yl benzoate (36)
General procedure was applied with alkyl trifluoroborate **S16b** (0.2 mmol), benzoic acid (0.6 mmol), and $^{n}Bu_4NPF_6$ (0.2 mmol). The reaction mixture was electrolyzed at a constant current of 5.0 mA for 2.0 h. Purification by flash column chromatography (silica gel, hexanes:EtOAc = 40:1) afforded **36** (26.2 mg, 51%) as a colorless oil; R_f = 0.7 (hexanes:EtOAc = 10:1); ^{1}H NMR (400 MHz, $CDCl_3$): δ 8.03–7.95 (m, 2H), 7.52 (t, J = 7.4 Hz, 1H), 7.41 (t, J = 7.7 Hz, 2H), 2.31–2.18 (m, 9H), 1.79–1.65 (m, 6H); ^{13}C NMR (101 MHz, $CDCl_3$): δ 165.6, 132.5, 132.3, 129.6, 128.3, 81.2, 41.6, 36.4, 31.1; HRMS (ESI) *m*/*z*: $[M+Na]^+$ calcd for $C_{17}H_{20}O_2Na$ 279.1361; found 279.1359.

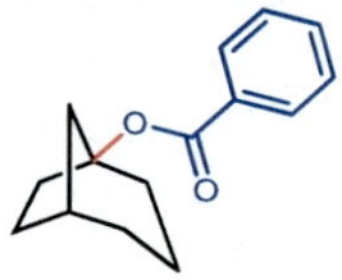

Bicyclo[3.2.1]octan-1-yl benzoate (37)
General procedure was applied with alkyl trifluoroborate **S11b** (0.2 mmol), benzoic acid (0.6 mmol), and $^{n}Bu_4NPF_6$ (0.2 mmol). The reaction mixture was electrolyzed at a constant current of 5.0 mA for 2.0 h. Purification by flash column chromatography (silica gel, hexanes:EtOAc = 50:1) afforded **37** (15.4 mg, 33%) as a colorless oil; R_f = 0.8 (hexanes:EtOAc = 10:1); ^{1}H NMR (400 MHz, $CDCl_3$): δ 8.00 (d, J = 7.4 Hz, 2H), 7.52 (t, J = 7.4 Hz, 1H), 7.46–7.37 (m, 2H), 2.37–2.17 (m, 3H), 2.08–2.00 (m, 1H), 1.99–1.82 (m, 4H), 1.75–1.62 (m, 2H), 1.52–1.38 (m, 3H); ^{13}C NMR (101 MHz, $CDCl_3$): δ 166.0, 132.6, 131.8, 129.6, 128.3, 88.1, 43.9, 35.7, 34.6, 33.8, 31.6, 27.6, 20.3; HRMS (ESI) *m*/*z*: $[M+Na]^+$ calcd for $C_{15}H_{18}O_2Na$ 253.1205; found 253.1202.

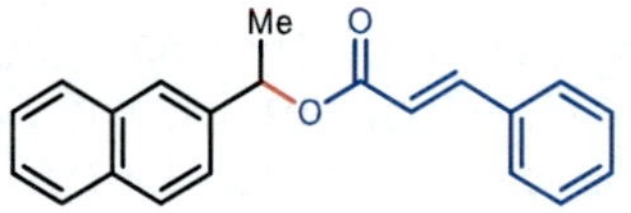

1-(Naphthalen-2-yl)ethyl cinnamate (38)
General procedure was applied with alkyl trifluoroborate **S5b** (0.2 mmol), cinnamic acid (0.6 mmol), and $^{n}Bu_4NClO_4$ (0.2 mmol). The reaction mixture was electrolyzed

at a constant current of 5.0 mA for 2.0 h. Purification by flash column chromatography (silica gel, hexanes:EtOAc = 30:1) afforded **38** (33.9 mg, 56%) as a colorless oil; R_f = 0.5 (hexanes:EtOAc = 10:1); ^{1}H NMR (400 MHz, $CDCl_3$) δ 7.89–7.81 (m, 4H), 7.72 (d, *J* = 16.0 Hz, 1H), 7.57–7.45 (m, 5H), 7.42–7.36 (m, 3H), 6.51 (d, *J* = 16.0 Hz, 1H), 6.20 (q, *J* = 6.6 Hz, 1H), 1.71 (d, *J* = 6.6 Hz, 3H); ^{13}C NMR (101 MHz, $CDCl_3$) δ 166.2, 144.9, 139.1, 134.4, 133.2, 133.0, 130.3, 128.9, 128.4, 128.1, 128.0, 127.6, 126.2, 126.0, 125.0, 124.1, 118.4, 72.5, 22.2; HRMS (ESI) *m/z*: [M+Na]$^+$ calcd for $C_{21}H_{18}O_2Na$ 325.1205; found 325.1201.

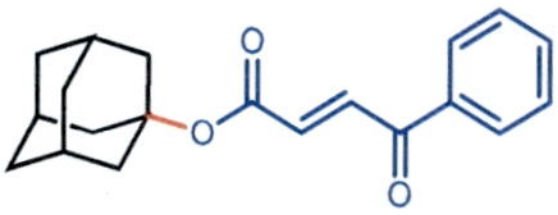

Adamantan-1-yl (*E*)-4-oxo-4-phenylbut-2-enoate (39)
General procedure was applied with alkyl trifluoroborate **S16b** (0.2 mmol), cinnamic acid (0.6 mmol), and $^nBu_4NPF_6$ (0.2 mmol). The reaction mixture was electrolyzed at a constant current of 5.0 mA for 2.0 h. Purification by flash column chromatography (silica gel, hexanes:EtOAc = 40:1) afforded **39** (50.5 mg, 81%) as a colorless oil; R_f = 0.6 (hexanes:EtOAc = 10:1); ^{1}H NMR (400 MHz, $CDCl_3$): δ 8.01–7.96 (m, 2H), 7.78 (d, *J* = 15.6 Hz, 1H), 7.61 (t, *J* = 7.4 Hz, 1H), 7.51 (t, *J* = 7.7 Hz, 2H), 6.80 (d, *J* = 15.6 Hz, 1H), 2.25–2.15 (m, 9H), 1.78–1.64 (m, 6H); ^{13}C NMR (101 MHz, $CDCl_3$): δ 190.2, 164.6, 136.9, 135.7, 135.0, 133.8, 129.0, 129.0, 82.2, 41.4, 36.3, 31.0; HRMS (ESI) *m/z*: [M+Na]$^+$ calcd for $C_{20}H_{22}O_3Na$ 333.1467; found 333.1465.

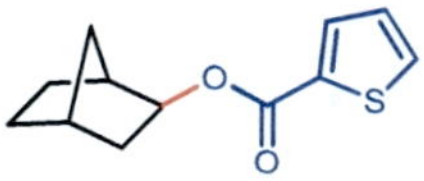

Bicyclo[2.2.1]heptan-2-yl thiophene-2-carboxylate (40)
General procedure was applied with alkyl trifluoroborate **S15b** (0.2 mmol), thiophene-2-carboxylic acid (0.6 mmol), and $^nBu_4NPF_6$ (0.2 mmol). The reaction mixture was electrolyzed at a constant current of 5.0 mA for 2.0 h. Purification by flash column chromatography (silica gel, hexanes:EtOAc = 40:1) afforded **40** (29.8 mg, 67%) as a colorless oil; R_f = 0.5 (hexanes:EtOAc = 10:1); ^{1}H NMR (500 MHz): δ 7.77 (dd, *J* = 3.7, 1.3 Hz, 1H), 7.52 (dd, *J* = 5.0, 1.3 Hz, 1H), 7.08 (dd, *J* = 5.0, 3.7 Hz, 1H), 4.82 (dt, *J* = 7.1, 1.3 Hz, 1H), 2.44–2.43 (m, 1H), 2.37–2.30 (m, 1H), 1.84–1.78 (m, 1H), 1.65–1.53 (m, 3H), 1.52–1.41 (m, 1H), 1.25–1.10 (m, 3H); ^{13}C NMR (101 MHz, $CDCl_3$): δ 161.9, 134.6, 133.0, 132.0, 127.6, 78.4, 41.5, 39.5, 35.4, 35.4, 28.2, 24.2; HRMS (ESI) *m/z*: [M+Na]$^+$ calcd for $C_{12}H_{14}O_2SNa$ 245.0612; found 245.0610.

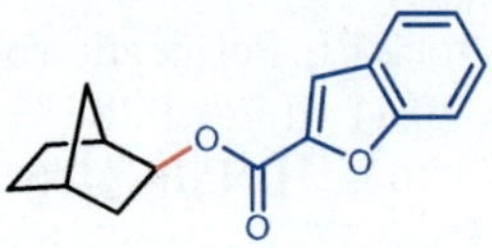

Bicyclo[2.2.1]heptan-2-yl benzofuran-2-carboxylate (41)

General procedure was applied with alkyl trifluoroborate **S15b** (0.2 mmol), benzofuran-2-carboxylic acid (0.6 mmol), and $^{n}Bu_4NPF_6$ (0.2 mmol). The reaction mixture was electrolyzed at a constant current of 5.0 mA for 2.0 h. Purification by flash column chromatography (silica gel, hexanes:EtOAc = 40:1) afforded **41** (25.5 mg, 50%) as a colorless oil; R_f = 0.5 (hexanes:EtOAc = 10:1); ^{1}H NMR (400 MHz, $CDCl_3$): δ 7.67 (d, J = 7.9 Hz, 1H), 7.59 (d, J = 8.4 Hz, 1H), 7.48 (d, J = 0.9 Hz, 1H), 7.47–7.40 (m, 1H), 7.32–7.27 (m, 1H), 4.91 (d, J = 7.1 Hz, 1H), 2.49 (d, J = 4.9 Hz, 1H), 2.36 (s, 1H), 1.89–1.83 (m, 1H), 1.70–1.55 (m, 3H), 1.55–1.44 (m, 1H), 1.29–1.15 (m, 3H); ^{13}C NMR (101 MHz, $CDCl_3$): δ 159.5, 155.8, 146.3, 127.6, 127.2, 123.8, 122.9, 113.6, 112.5, 79.0, 41.7, 39.7, 35.6, 35.6, 28.3, 24.4; HRMS (ESI) m/z: $[M+Na]^+$ calcd for $C_{16}H_{16}O_3Na$ 279.0997; found 279.0995.

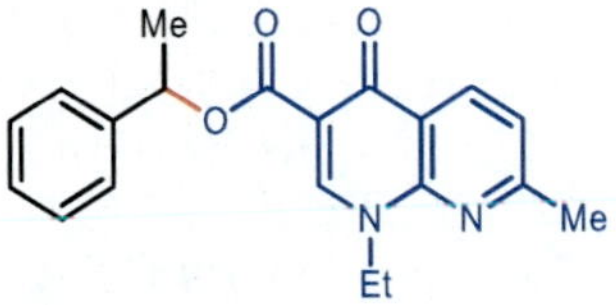

1-Phenylethyl 1-ethyl-7-methyl-4-oxo-1,4-dihydro-1,8-naphthyridine-3-carboxylate (42)

General procedure was applied with alkyl trifluoroborate **S1b** (0.2 mmol), 1-ethyl-7-methyl-4-oxo-1,4-dihydro-1,8-naphthyridine-3-carboxylic acid (0.6 mmol), and $^{n}Bu_4ClO_4$ (0.2 mmol). The reaction mixture was electrolyzed at a constant current of 5.0 mA for 2.0 h. Purification by flash column chromatography (silica gel, hexanes:EtOAc = 2:1) afforded **42** (37.0 mg, 55%) as a colorless oil; R_f = 0.2 (hexanes:EtOAc = 1:1); ^{1}H NMR (500 MHz, $CDCl_3$): δ 8.66 (d, J = 7.7 Hz, 1H), 8.64 (s, 1H), 7.54 (d, J = 7.5 Hz, 2H), 7.36 (t, J = 7.5 Hz, 2H), 7.31–7.23 (m, 2H), 6.15 (q, J = 6.6 Hz, 1H), 4.49–4.46 (m, 2H), 2.66 (s, 3H), 1.69 (d, J = 6.0 Hz, 3H), 1.48 (t, J = 7.2 Hz, 3H); ^{13}C NMR (126 MHz, $CDCl_3$): δ 174.7, 165.2, 162.9, 149.0, 148.8, 142.3, 137.0, 128.6, 127.8, 126.3, 121.6, 121.33, 112.1, 73.2, 46.8, 25.2, 22.9, 15.3; HRMS (ESI) m/z: $[M+H]^+$ calcd for $C_{20}H_{21}N_2O_3$ 337.1552; found 337.1548.

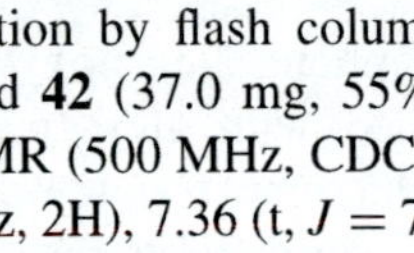

1-Phenylethyl tetrahydrofuran-2-carboxylate (43)

General procedure was applied with alkyl trifluoroborate **S1b** (0.2 mmol), tetrahydrofuran-2-carboxylic acid (0.6 mmol), and $^{n}Bu_4NClO_4$ (0.2 mmol). The reaction mixture was electrolyzed at a constant current of 5.0 mA for 2.0 h. Purification by flash column chromatography (silica gel, hexanes:EtOAc = 10:1) afforded **43** (30.0 mg, 68%) as a colorless oil; $R_f = 0.5$ (hexanes:EtOAc = 3:1); ^{1}H NMR (400 MHz, $CDCl_3$, for both diastereomers) (the integration at 1.56 ppm and 1.56 ppm indicated the ratio of the two isomers of **29** to be 1:1): δ 7.39–7.31 (m, 4H), 7.31–7.28 (m, 1H), 5.97–5.91 (m, 1H), 4.48 (dt, $J = 8.6, 5.7$ Hz, 1H), 4.07–3.97 (m, 1H), 3.98–3.89 (m, 1H), 2.33–2.17 (m, 1H), 2.08–1.84 (m, 3H), 1.56 (d, $J = 6.6$ Hz, 1.5H), 1.56 (d, $J = 6.6$ Hz, 1.5H); ^{13}C NMR (101 MHz, $CDCl_3$, for both diastereomers): δ 172.8, 172.8, 141.5, 141.3, 128.6, 128.1, 128.0, 126.2, 126.1, 77.0, 72.9, 72.8, 69.5, 30.3, 30.2, 25.3, 25.3, 22.4, 22.2; HRMS (ESI) *m/z*: $[M+Na]^+$ calcd for $C_{13}H_{16}O_3Na$ 243.0997; found 243.0994.

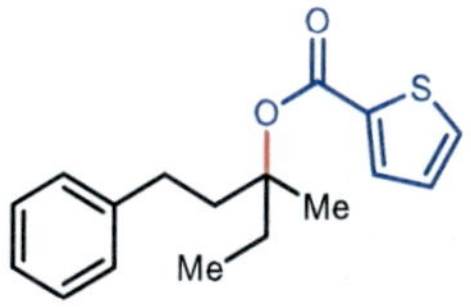

3-Methyl-1-phenylpentan-3-yl thiophene-2-carboxylate (44)

General procedure was applied with alkyl trifluoroborate **S24b** (0.2 mmol), thiophene-2-carboxylic acid (0.6 mmol), and $^{n}Bu_4NClO_4$ (0.2 mmol). The reaction mixture was electrolyzed at a constant current of 5.0 mA for 2.0 h. Purification by flash column chromatography (silica gel, hexanes:EtOAc = 30:1) afforded **44** (15.7 mg, 27%) as a colorless oil; $R_f = 0.5$ (hexanes:EtOAc = 10:1); ^{1}H NMR (400 MHz, $CDCl_3$) δ 7.74 (d, $J = 3.7$ Hz, 1H), 7.51 (d, $J = 5.0$ Hz, 1H), 7.27 (d, $J = 7.5$ Hz, 2H), 7.24–7.15 (m, 3H), 7.10–7.06 (m, 1H), 2.77–2.63 (m, 2H), 2.36–2.20 (m, 1H), 2.20–2.03 (m, 2H), 1.98–1.92 (m, 1H), 1.61 (s, 3H), 0.97 (t, $J = 7.5$ Hz, 3H); ^{13}C NMR (126 MHz, $CDCl_3$) δ 161.5, 142.3, 135.8, 132.9, 131.9, 128.6, 128.5, 127.8, 126.0, 86.4, 40.5, 31.4, 30.3, 23.6, 8.3; HRMS (ESI) *m/z*: $[M+Na]^+$ calcd for $C_{17}H_{20}O_2SNa$ 311.1082; found 311.1080.

1-(2,4,6-Trimethoxyphenyl)adamantane (45)

General procedure was applied with alkyl trifluoroborate **S16b** (0.2 mmol), 1,3,5-trimethoxybenzene (0.6 mmol), and $^{n}Bu_4NPF_6$ (0.2 mmol). Purification by flash column chromatography (silica gel, petroleum ether:EtOAc = 30:1) afforded **45**

(21.2 mg, 35%) as a white solid; m.p. 113–115 °C; $R_f = 0.6$ (petroleum ether:EtOAc = 7:1); ^{1}H NMR (400 MHz, $CDCl_3$): δ 6.14 (s, 2H), 3.78 (s, 3H), 3.75 (s, 6H), 2.37–2.25 (m, 6H), 2.05–1.93 (m, 3H), 1.85–1.66 (m, 6H); ^{13}C NMR (101 MHz, $CDCl_3$): δ 160.8, 158.6, 119.3, 93.6, 56.2, 55.2, 41.7, 39.8, 37.5, 29.8; HRMS (ESI) *m*/*z*: $[M+H]^+$ calcd for $C_{19}H_{27}O_3$ 303.1960; found 303.1956.

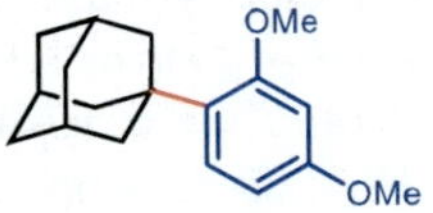

1-(2,4-Dimethoxyphenyl)adamantane (46)

General procedure was applied with alkyl trifluoroborate **S16b** (0.2 mmol), 1,3-dimethoxybenzene (0.6 mmol), and ^{n}Bu$_4$NPF$_6$ (0.2 mmol). Purification by flash column chromatography (silica gel, hexanes:CH_2Cl_2 = 8:1) afforded **46** (22.5 mg, 41%) as a white solid; m.p. 100–102 °C; $R_f = 0.7$ (hexanes:CH_2Cl_2 = 2:1); ^{1}H NMR (400 MHz, $CDCl_3$): δ 7.11 (d, *J* = 8.5 Hz, 1H), 6.49–6.41 (m, 2H), 3.81 (s, 3H), 3.79 (s, 3H), 2.10–2.00 (m, 9H), 1.82–1.71 (m, 6H); ^{13}C NMR (101 MHz, $CDCl_3$): δ 159.9, 158.9, 131.5, 126.9, 103.6, 99.9, 55.4, 55.1, 41.0, 37.3, 36.5, 29.3; HRMS (ESI) *m*/*z*: $[M+Na]^+$ calcd for $C_{18}H_{24}O_2Na$ 295.1674; found 295.1676.

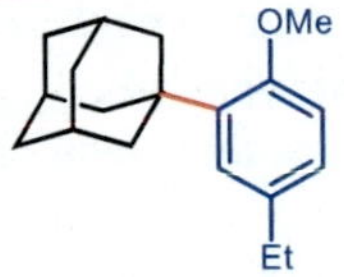

1-(5-Ethyl-2-methoxyphenyl)adamantane (47)

General procedure was applied with alkyl trifluoroborate **S16b** (0.2 mmol), 1-ethyl-4-methoxybenzene (0.6 mmol), and ^{n}Bu$_4$NPF$_6$ (0.2 mmol). Purification by flash column chromatography (silica gel, hexanes:CH_2Cl_2 = 12:1) afforded **47** (18.0 mg, 33%) as a colorless oil; $R_f = 0.8$ (hexanes:CH_2Cl_2 = 3:1); ^{1}H NMR (400 MHz, $CDCl_3$): δ 7.05 (d, *J* = 2.2 Hz, 1H), 7.00 (dd, *J* = 8.2, 2.3 Hz, 1H), 6.81 (d, *J* = 8.2 Hz, 1H), 3.81 (s, 3H), 2.59 (q, *J* = 7.6 Hz, 2H), 2.15–2.01 (m, 9H), 1.83–1.72 (m, 6H), 1.22 (t, *J* = 7.6 Hz, 3H); ^{13}C NMR (101 MHz, $CDCl_3$): δ 157.0, 138.4, 136.0, 126.4, 125.7, 111.8, 55.3, 40.8, 37.3, 37.1, 29.3, 28.5, 16.0; HRMS (EI) *m*/*z*: $[M]^+$ calcd for $C_{19}H_{26}O$ 270.1984; found 270.1985.

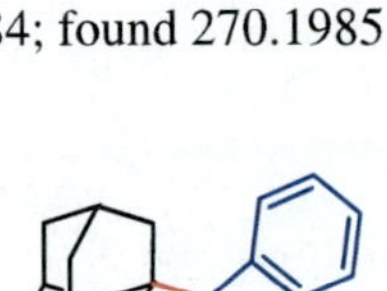

3-(-Adamantan-1-yl)-1-tosyl-1*H*-indole (48)
General procedure was applied with alkyl trifluoroborate **S16b** (0.1 mmol), 1-tosyl-1*H*-indole (0.2 mmol), and $^{n}Bu_4NPF_6$ (0.2 mmol). The reaction mixture was electrolyzed at a constant current of 3.0 mA for 3.0 h. Purification by flash column chromatography (silica gel, hexanes:EtOAc = 19:1) afforded **48** (16.7 mg, 41%) as a white solid; m.p. 178–180 °C; R_f = 0.5 (hexanes:EtOAc = 9:1); ^{1}H NMR (400 MHz, $CDCl_3$): 7.98 (d, J = 8.2 Hz, 1H), 7.79 (d, J = 7.9 Hz, 1H), 7.74 (d, J = 8.1 Hz, 2H), 7.30–7.15 (m, 5H), 2.34 (s, 3H), 2.15–2.00 (m, 9H), 1.88–1.75 (m, 6H); ^{13}C NMR (101 MHz, $CDCl_3$): δ 144.7, 136.1, 135.6, 133.0, 129.9, 129.5, 126.9, 124.0, 122.5, 122.2, 121.2, 114.1, 42.3, 37.1, 34.3, 28.7, 21.7; HRMS (ESI) *m*/*z*: $[M+Na]^+$ calcd for $C_{25}H_{27}NO_2SNa$ 428.1660; found 428.1657.

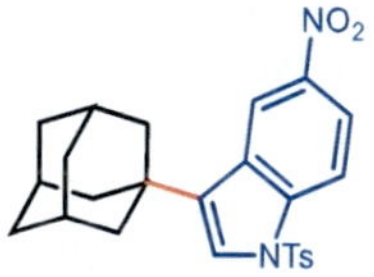

3-(Adamantan-1-yl)-5-nitro-1-tosyl-1*H*-indole (49)
General procedure was applied with alkyl trifluoroborate **S16b** (0.1 mmol), 5-nitro-1-tosyl-1*H*-indole (0.2 mmol), and $^{n}Bu_4NPF_6$ (0.2 mmol). The reaction mixture was electrolyzed at a constant current of 3.0 mA for 3.0 h. Purification by flash column chromatography (silica gel, hexanes:EtOAc = 19:1) afforded **49** (15.3 mg, 34%) as a white solid; m.p. 181–183 °C; R_f = 0.7 (hexanes:EtOAc = 4:1); ^{1}H NMR (400 MHz, $CDCl_3$): δ 8.68 (s, 1H), 8.16 (d, J = 9.2 Hz, 1H), 8.06 (d, J = 9.2 Hz, 1H), 7.77 (d, J = 8.1 Hz, 2H), 7.40 (s, 1H), 7.27 (d, J = 8.1 Hz, 2H), 2.37 (s, 3H), 2.17–2.09 (m, 3H), 2.08–2.01 (m, 6H), 1.90–1.77 (m, 6H); ^{13}C NMR (101 MHz, $CDCl_3$): δ 145.8, 143.4, 139.0, 135.0, 133.6, 130.3, 129.1, 127.0, 124.0, 119.4, 118.5, 114.0, 42.5, 36.9, 34.3, 28.6, 21.8; HRMS (ESI) *m*/*z*: $[M-H]^-$ calcd for $C_{25}H_{25}N_2O_4S$ 449.1541; found 449.1530.

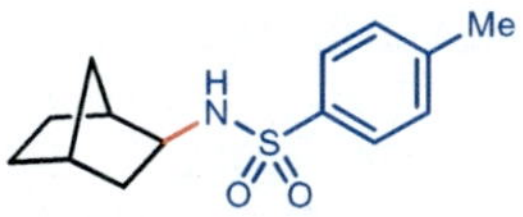

***N*-(Bicyclo[2.2.1]heptan-2-yl)-4-methylbenzenesulfonamide (50)**
General procedure was applied with alkyl trifluoroborate **S15b** (0.2 mmol), 4-methylbenzenesulfonamide (0.6 mmol), and $^{n}Bu_4NPF_6$ (0.2 mmol). Purification by flash column chromatography (silica gel, hexanes:EtOAc = 10:1) afforded **50** (32.4 mg, 61%) as a colorless oil; R_f = 0.5 (hexanes:EtOAc = 3:1); ^{1}H NMR (400 MHz, $CDCl_3$): δ 7.75 (d, J = 8.2 Hz, 2H), 7.30 (d, J = 8.0 Hz, 2H), 4.53 (d, J = 6.5 Hz, 1H), 3.13 (s, 1H), 2.43 (s, 3H), 2.19 (s, 1H), 2.10 (d, J = 4.1 Hz,

1H), 1.62–1.56 (m, 1H), 1.43–1.37 (m, 2H), 1.33–1.30 (m, 1H), 1.20–1.08 (m, 2H), 1.08–0.93 (m, 2H); ^{13}C NMR (101 MHz, $CDCl_3$): δ 143.4, 138.1, 129.8, 127.2, 56.8, 42.6, 41.0, 35.7, 35.3, 28.1, 26.5, 21.7; HRMS (ESI) *m*/*z*: $[M+Na]^+$ calcd for $C_{14}H_{19}NO_2SNa$ 288.1034; found 288.1032.

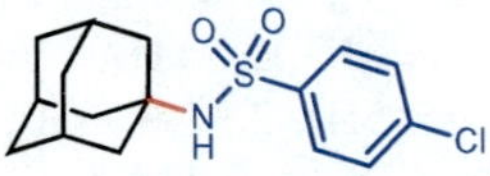

N-(Adamantan-1-yl)-4-chlorobenzenesulfonamide (51)
General procedure was applied with alkyl trifluoroborate **S16b** (0.2 mmol), 4-chlorobenzenesulfonamide (0.6 mmol), and $^nBu_4NPF_6$ (0.2 mmol). Purification by flash column chromatography (silica gel, hexanes:EtOAc = 10:1) afforded **51** (50.3 mg, 77%) as a white solid; m.p. 187–189 °C; R_f = 0.5 (hexanes:EtOAc = 3:1); ^{1}H NMR (400 MHz): δ 7.84 (d, *J* = 8.6 Hz, 2H), 7.45 (d, *J* = 8.6 Hz, 2H), 4.82 (s, 1H), 2.05–1.95 (m, 3H), 1.83–1.73 (m, 6H), 1.66–1.50 (m, 6H); ^{13}C NMR (101 MHz, $CDCl_3$): δ 142.7, 138.6, 129.3, 128.5, 55.5, 43.2, 35.9, 29.6; HRMS (ESI) *m*/*z*: $[M+Na]^+$ calcd for $C_{16}H_{20}ClNO_2SNa$ 348.0801; found 348.0793.

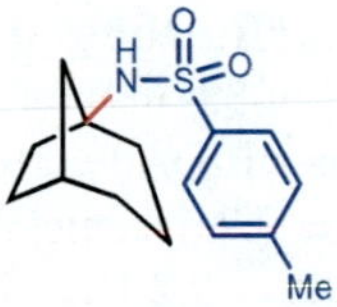

N-(Bicyclo[3.2.1]octan-1-yl)-4-methylbenzenesulfonamide (52))
General procedure was applied with alkyl trifluoroborate **S11b** (0.2 mmol), 4-methylbenzenesulfonamide (0.6 mmol), and $^nBu_4NPF_6$ (0.2 mmol). Purification by flash column chromatography (silica gel, hexanes:EtOAc = 6:1) afforded **52** (15.2 mg, 27%) as a colorless oil; R_f = 0.5 (hexanes:EtOAc = 2:1); ^{1}H NMR (400 MHz, $CDCl_3$): δ 7.76 (d, *J* = 8.1 Hz, 2H), 7.27 (d, *J* = 7.8 Hz, 2H), 4.69 (s, 1H), 2.42 (s, 3H), 2.18–2.17 (m, 1H), 1.77–1.65 (m, 5H), 1.56–1.31 (m, 5H), 1.30–1.18 (m, 2H); ^{13}C NMR (101 MHz, $CDCl_3$): δ 143.0, 140.6, 129.6, 127.1, 63.4, 44.5, 38.8, 35.5, 34.9, 31.2, 28.0, 21.7, 19.8; HRMS (ESI) *m*/*z*: $[M+Na]^+$ calcd for $C_{15}H_{21}NO_2SNa$ 302.1191; found 302.1182.

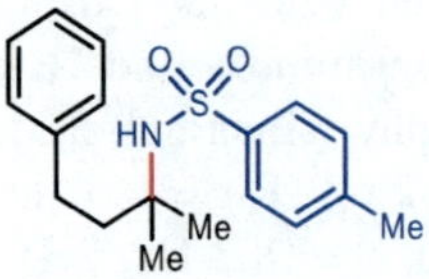

4-Methyl-*N*-(2-methyl-4-phenylbutan-2-yl)benzenesulfonamide (53)

General procedure was applied with alkyl trifluoroborate **S22b** (0.2 mmol), 4-methylbenzenesulfonamide (0.6 mmol), and $^{n}Bu_4NPF_6$ (0.2 mmol). Purification by flash column chromatography (silica gel, hexanes:EtOAc = 6:1) afforded **53** (32.0 mg, 50%) as a colorless oil; R_f = 0.5 (hexanes:EtOAc = 2:1); ^{1}H NMR (400 MHz, $CDCl_3$) δ 7.83–7.72 (m, 2H), 7.30–7.22 (m, 4H), 7.22–7.14 (m, 1H), 7.14–7.07 (m, 2H), 4.56 (s, 1H), 2.66–2.53 (m, 2H), 2.40 (s, 3H), 1.84–1.73 (m, 2H), 1.22 (s, 6H); ^{13}C NMR (101 MHz, $CDCl_3$) δ 143.1, 141.8, 140.6, 129.7, 128.6, 128.5, 127.2, 126.0, 57.2, 45.0, 30.5, 28.0, 21.6; HRMS (ESI) *m*/*z*: $[M+Na]^+$ calcd for $C_{18}H_{23}NO_2SNa$ 340.1347; found 340.1344.

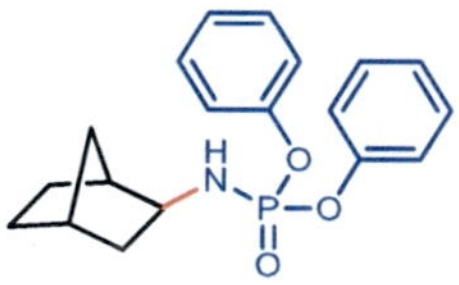

Diphenyl bicyclo[2.2.1]heptan-2-ylphosphoramidate (54)

General procedure was applied with alkyl trifluoroborate **S15b** (0.2 mmol), diphenyl phosphoramidate (0.6 mmol), and $^{n}Bu_4NPF_6$ (0.2 mmol). Purification by flash column chromatography (silica gel, hexanes:EtOAc = 6:1) afforded **54** (29.7 mg, 43%) as a colorless oil; R_f = 0.2 (hexanes:EtOAc = 4:1); ^{1}H NMR (400 MHz, $CDCl_3$): δ 7.33–7.29 (m, 4H), 7.28–7.19 (m, 4H), 7.14 (t, *J* = 7.3 Hz, 2H), 3.27 (q, *J* = 8.5 Hz, 1H), 2.95–2.89 (m, 1H), 2.25–2.11 (m, 2H), 1.79–1.70 (m, 1H), 1.52–1.35 (m, 2H), 1.28–1.24 (m, 1H), 1.20–1.03 (m, 4H); ^{13}C NMR (101 MHz, $CDCl_3$): δ 129.8, 125.0, 120.5, 120.4, 120.4, 55.4, 43.9, 43.8, 42.7, 42.6, 35.8, 35.1, 28.1, 26.6; ^{31}P NMR (243 MHz, $CDCl_3$): δ −2.73; HRMS (ESI) *m*/*z*: $[M+H]^+$ calcd for $C_{19}H_{23}NO_3P$ 344.1416; found 344.1410.

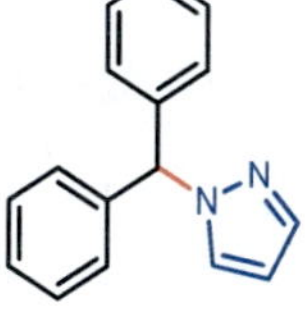

1-Benzhydryl-1*H*-pyrazole (55)

General procedure was applied with alkyl trifluoroborate **S4b** (0.2 mmol), 1*H*-pyrazole (0.6 mmol), and $^{n}Bu_4NClO_4$ (0.2 mmol). Purification by flash column chromatography (silica gel, hexanes:EtOAc = 20:1) afforded **55** (45.0 mg, 96%) as a colorless oil; R_f = 0.3 (hexanes:EtOAc = 10:1); ^{1}H NMR (400 MHz, $CDCl_3$): δ 7.60 (s, 1H), 7.33 (d, *J* = 7.6 Hz, 7H), 7.15–7.04 (m, 4H), 6.79 (s, 1H), 6.27 (s, 1H);

^{13}C NMR (101 MHz, $CDCl_3$): δ 139.9, 139.7, 129.6, 128.8, 128.4, 128.2, 105.5, 69.6; HRMS (ESI) *m*/*z*: $[M+H]^+$ calcd for $C_{16}H_{15}N_2$ 235.1235; found 235.1230.

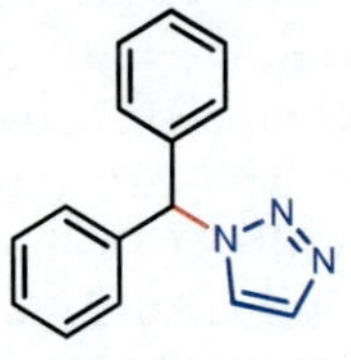

1-Benzhydryl-1*H*-1,2,3-triazole (56)

General procedure was applied with alkyl trifluoroborate **S4b** (0.2 mmol), 1*H*-1,2,3-triazole (0.6 mmol), and $^n Bu_4NClO_4$ (0.2 mmol). Purification by flash column chromatography (silica gel, hexanes:EtOAc = 8:1) afforded **56** (30.1 mg, 64%) as a white solid; m.p. 121–123 °C; $R_f = 0.3$ (hexanes:EtOAc = 3:1); ^{1}H NMR (400 MHz, $CDCl_3$): δ 7.72 (s, 1H), 7.43 (s, 1H), 7.37–7.35 (m, 6H), 7.16–7.07 (m, 5H); ^{13}C NMR (101 MHz, $CDCl_3$): δ 138.2, 133.6, 133.6, 128.9, 128.9, 128.6, 128.1, 123.6, 123.5, 67.9; HRMS (ESI) *m*/*z*: $[M+H]^+$ calcd for $C_{15}H_{14}N_3$ 236.1188; found 236.1183.

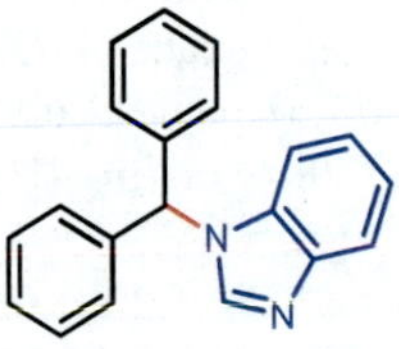

1-Benzhydryl-1*H*-benzo[*d*]imidazole (57)

General procedure was applied with alkyl trifluoroborate **S4b** (0.2 mmol), 1*H*-benzo[*d*]imidazole (0.6 mmol), and $^n Bu_4NClO_4$ (0.2 mmol). Purification by flash column chromatography (silica gel, hexanes:EtOAc = 4:1) afforded **57** (50.8 mg, 89%) as a white solid; m.p. 158–160 °C; $R_f = 0.3$ (hexanes:EtOAc = 2:1); ^{1}H NMR (400 MHz, $CDCl_3$): δ 7.83 (d, *J* = 8.0 Hz, 1H), 7.61 (s, 1H), 7.36–7.10 (m, 13H), 6.75 (s, 1H); ^{13}C NMR (101 MHz, $CDCl_3$): δ 144.3, 142.7, 138.2, 134,2, 129.2, 128.7, 128.4, 123.1, 122.5, 120.6, 110.9, 63.8; HRMS (ESI) *m*/*z*: $[M+H]^+$ calcd for $C_{20}H_{17}N_2$ 285.1392; found 285.1383.

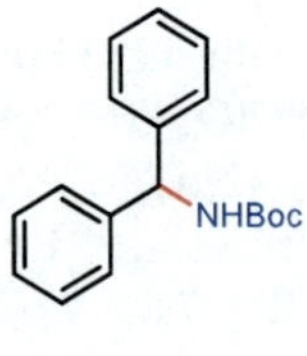

***tert*-butyl benzhydrylcarbamate (58)**

General procedure was applied with alkyl trifluoroborate **S4b** (0.2 mmol), *tert*-butyl carbamate (0.6 mmol), and $^{n}Bu_4NClO_4$ (0.2 mmol). Purification by flash column chromatography (silica gel, hexanes:EtOAc = 30:1) afforded **58** (36.4 mg, 64%) as a white solid; m.p. 123–125 °C; R_f = 0.4 (hexanes:EtOAc = 10:1); Spectroscopic data matches with previously reported data [40].

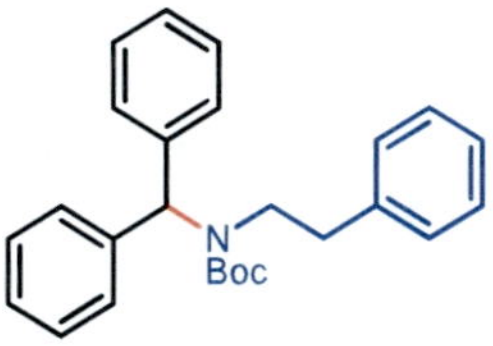

***tert*-butyl benzhydryl(phenethyl)carbamate (59)**

General procedure was applied with alkyl trifluoroborate **S4b** (0.2 mmol), *tert*-butyl phenethylcarbamate (0.6 mmol), and $^{n}Bu_4NClO_4$ (0.2 mmol). Purification by flash column chromatography (silica gel, hexanes:EtOAc = 30:1) afforded **59** (38.1 mg, 49%) as a colorless oil; R_f = 0.5 (hexanes:EtOAc = 10:1); ^{1}H NMR (400 MHz, $CDCl_3$): δ 7.37–7.28 (m, 6H), 7.25–7.21 (m, 4H), 7.18–7.08 (m, 3H), 6.86–6.60 (m, 3H), 3.43–3.30 (m, 2H), 2.30–2.06 (m, 2H), 1.59–1.38 (m, 9H); ^{13}C NMR (101 MHz, $CDCl_3$): δ 155.9, 140.4, 139.5, 129.0, 128.8, 128.5, 128.4, 127.5, 126.2, 80.2, 77.4, 47.8, 35.5, 28.6; HRMS (ESI) *m/z*: $[M+Na]^+$ calcd for $C_{26}H_{29}NO_2Na$ 410.2096; found 410.2094.

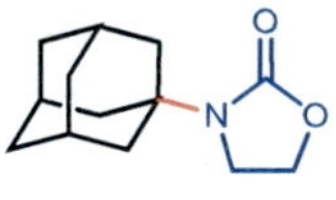

3-(Adamantan-1-yl)oxazolidin-2-one (60)

General procedure was applied with alkyl trifluoroborate **S16b** (0.2 mmol), oxazolidin-2-one (0.6 mmol), and $^{n}Bu_4NPF_6$ (0.2 mmol). Purification by flash column chromatography (silica gel, hexanes:EtOAc = 1:1) afforded **60** (19.3 mg, 44%) as a white solid; m.p. 127–129 °C; R_f = 0.1 (hexanes:EtOAc = 2:1); ^{1}H NMR (400 MHz, $CDCl_3$): δ 4.24–4.17 (m, 2H), 3.63–3.56 (m, 2H), 2.16–2.00 (m, 9H), 1.75–1.59 (m, 6H); ^{13}C NMR (101 MHz, $CDCl_3$): δ 157.0, 61.4, 54.2, 42.5, 39.6, 36.3, 29.6; HRMS (ESI) *m/z*: $[M+H]^+$ calcd for $C_{13}H_{20}NO_2$ 222.1494; found 222.1488.

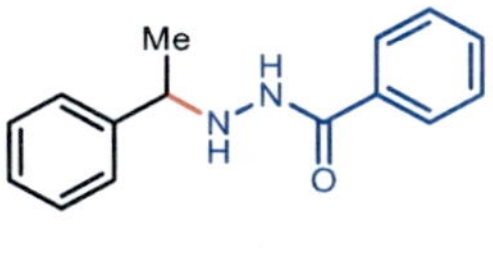

N'-(1-Phenylethyl)benzohydrazide (61)

General procedure was applied with alkyl trifluoroborate **S1b** (0.2 mmol), benzohydrazide (0.6 mmol), and $^{n}Bu_4NClO_4$ (0.2 mmol). Purification by flash column chromatography (silica gel, hexanes:EtOAc = 9:1) afforded **61** (27.0 mg, 56%) as a colorless oil; $R_f = 0.4$ (hexanes:EtOAc = 3:1); ^{1}H NMR (500 MHz, $CDCl_3$): δ 7.64–7.60 (m, 2H), 7.51–7.45 (m, 2H), 7.43–7.33 (m, 5H), 7.32–7.27 (m, 1H), 5.10 (s, 1H), 4.26 (q, $J = 6.6$ Hz, 1H), 1.66 (s, 1H), 1.43 (d, $J = 6.6$ Hz, 3H); ^{13}C NMR (126 MHz, $CDCl_3$): δ 167.5, 143.2, 133.0, 131.9, 128.8, 128.8, 127.7, 127.4, 127.0, 60.2, 21.4; HRMS (ESI) *m/z*: $[M+H]^+$ calcd for $C_{15}H_{17}N_2O$ 241.1341; found 241.1334.

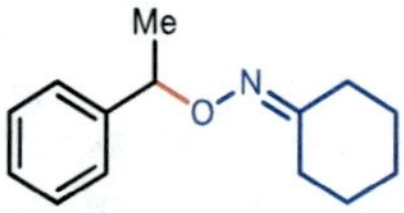

Cyclohexanone O-(1-phenylethyl) oxime (62)

General procedure was applied with alkyl trifluoroborate **S1b** (0.2 mmol), cyclohexanone oxime (0.6 mmol), and $^{n}Bu_4NClO_4$ (0.2 mmol). Purification by flash column chromatography (silica gel, hexanes:CH_2Cl_2 = 5:1) afforded **62** (24.4 mg, 56%) as a colorless oil; $R_f = 0.4$ (hexanes:CH_2Cl_2 = 2:1); ^{1}H NMR (500 MHz, $CDCl_3$): δ 7.39–7.30 (m, 4H), 7.27–7.23 (m, 1H), 5.20 (q, $J = 6.6$ Hz, 1H), 2.56–2.54 (m, 2H), 2.25–2.07 (m, 2H), 1.68–1.56 (m, 6H), 1.52 (d, $J = 6.6$ Hz, 3H); ^{13}C NMR (101 MHz, $CDCl_3$): δ 160.7, 144.3, 128.4, 127.2, 126.2, 80.1, 32.4, 27.2, 26.0, 26.0, 25.7, 22.6; HRMS (ESI) *m/z*: $[M+H]^+$ calcd for $C_{14}H_{20}NO$ 218.1545; found 218.1538.

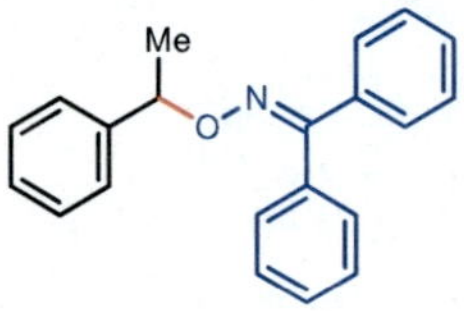

Diphenylmethanone O-(1-phenylethyl) oxime (63)

General procedure was applied with alkyl trifluoroborate **S1b** (0.2 mmol), diphenylmethanone oxime (0.6 mmol), and $^{n}Bu_4NClO_4$ (0.2 mmol). Purification by flash column chromatography (silica gel, hexanes:EtOAc = 7:1) afforded **63** (26.0 mg, 43%) as a colorless oil; $R_f = 0.7$ (hexanes:CH_2Cl_2 = 2:1); ^{1}H NMR (400 MHz, $CDCl_3$): δ 7.47–7.20 (m, 15H), 5.39 (q, $J = 6.6$ Hz, 1H), 1.54 (d, $J = 6.6$ Hz, 3H); ^{13}C NMR (101 MHz, $CDCl_3$): δ 156.7, 143.7, 136.9, 133.7, 129.6, 129.2, 128.8, 128.3, 128.3, 128.1, 128.1, 127.3, 126.4, 81.8, 22.4; HRMS (ESI) *m/z*: $[M+H]^+$ calcd for $C_{21}H_{20}NO$ 302.1545; found 302.1538.

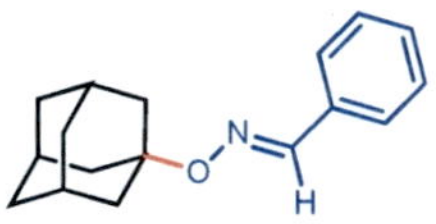

(*E*)-benzaldehyde *O*-(adamantan-1-yl) oxime (64)

General procedure was applied with alkyl trifluoroborate **S16b** (0.2 mmol), (*E*)-benzaldehyde oxime (0.6 mmol), and $^{n}Bu_4NPF_6$ (0.2 mmol). Purification by flash column chromatography (silica gel, hexanes:CH_2Cl_2 = 6:1)) afforded **64** (24.1 mg, 47%) as a colorless oil; R_f = 0.6 (hexanes:CH_2Cl_2 = 2:1); ^{1}H NMR (400 MHz, $CDCl_3$): δ 8.08 (s, 1H), 7.60–7.58 (m, 2H), 7.45–7.30 (m, 3H), 2.26–2.15 (m, 3H), 2.00–1.90 (m, 6H), 1.74–1.64 (m, 6H); ^{13}C NMR (101 MHz, $CDCl_3$): δ 147.6, 133.4, 129.4, 128.7, 127.0, 78.4, 41.8, 36.6, 30.8; HRMS (ESI) *m*/*z*: $[M+H]^+$ calcd for $C_{17}H_{22}NO$ 256.1701; found 256.1694.

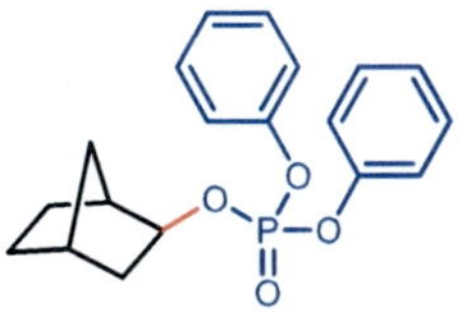

Bicyclo[2.2.1]heptan-2-yl diphenyl phosphate (65)

General procedure was applied with alkyl trifluoroborate **S15b** (0.2 mmol), diphenyl hydrogen phosphate (0.6 mmol), and $^{n}Bu_4NPF_6$ (0.2 mmol). Purification by flash column chromatography (silica gel, hexanes:EtOAc = 4:1) afforded **65** (29.3 mg, 43%) as a colorless oil; R_f = 0.33 (hexanes:EtOAc = 4:1); ^{1}H NMR (500 MHz, $CDCl_3$): δ 7.38–7.29 (m, 4H), 7.25–7.15 (m, 6H), 4.59 (t, *J* = 6.1 Hz, 1H), 2.45 (d, *J* = 5.0 Hz, 1H), 2.28 (s, 1H), 1.77–1.68 (m, 1H), 1.66–1.48 (m, 3H), 1.46–1.37 (m, 1H), 1.18–1.13 (m, 1H), 1.10–1.00 (m, 2H); ^{13}C NMR (126 MHz, $CDCl_3$): δ 150.8, 150.8, 129.8, 125.3, 120.3, 120.3, 120.2, 120.2, 83.4, 83.4, 42.9, 42.9, 40.6, 40.5, 35.5, 35.0, 28.1, 23.9; ^{31}P NMR (243 MHz, $CDCl_3$): δ −13.63; HRMS (ESI) *m*/*z*: $[M+Na]^+$ calcd for $C_{19}H_{21}O_4PNa$ 367.1075; found 367.1074.

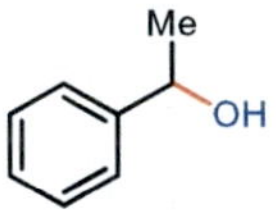

1-Phenylethan-1-ol (66)

General procedure was applied with alkyl trifluoroborate **S1b** (0.2 mmol), H_2O (0.1 mL), and $^{n}Bu_4NClO_4$ (0.2 mmol). The reaction mixture was electrolyzed at a constant current of 5.0 mA for 1.5 h. Purification by flash column chromatography (silica gel, hexanes:EtOAc = 7:3) afforded **66** (15.8 mg, 65%) as a colorless oil; R_f

= 0.2 (hexanes:EtOAc = 4:1) Spectroscopic data matches with previously reported data [41].

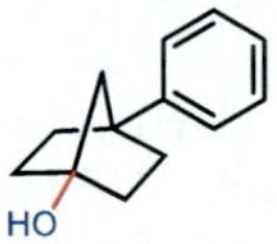

4-Phenylbicyclo[2.2.1]heptan-1-ol (67)
General procedure was applied with alkyl trifluoroborate **S14b** (0.2 mmol), H_2O (0.1 mL), and $^nBu_4NPF_6$ (0.2 mmol). The reaction mixture was electrolyzed at a constant current of 5.0 mA for 3.0 h. Purification by flash column chromatography (silica gel, hexanes:EtOAc = 2:1) afforded **67** (15.5 mg, 41%) as a colorless oil; R_f = 0.4 (hexanes:EtOAc = 1:1); ^{1}H NMR (400 MHz, $CDCl_3$) δ 7.41–7.13 (m, 5H), 2.03–1.68 (m, 10H); ^{13}C NMR (101 MHz, $CDCl_3$) δ 146.2, 128.4, 126.1, 125.9, 82.2, 48.6, 48.1, 37.5, 36.7; HRMS (ESI) *m*/*z*: $[M+Na]^+$ calcd for $C_{13}H_{16}ONa$ 211.1099; found 211.1094.

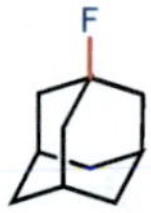

1-Fluoroadamantane (68)
General procedure was applied with alkyl trifluoroborate **S16b** (0.2 mmol), potassium hydrogen difluoride (0.4 mmol), 1,4,7,10,13,16-hexaoxacyclooctadecane (0.4 mmol), and $^nBu_4NPF_6$ (0.2 mmol). Purification by PTLC (silica gel, 100% hexanes) afforded **68** (21.1 mg, 68%) as a white solid; R_f = 0.5 (hexanes). Spectroscopic data matches with previously reported data [38]. Note: for **68**, the solvent was removed under reduced pressure (120 mbar) at 30 °C and the use of a high vacuum pump was avoided during the whole process due to the volatility of the product.

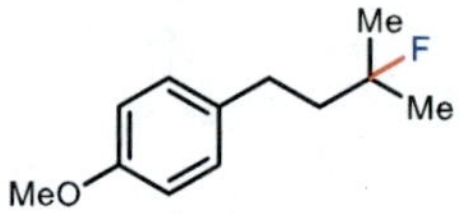

1-(3-Fluoro-3-methylbutyl)-4-methoxybenzene (69)
General procedure was applied with alkyl trifluoroborate **S23b** (0.2 mmol), potassium hydrogen difluoride (0.4 mmol), 1,4,7,10,13,16-hexaoxacyclooctadecane (0.4 mmol), and $^nBu_4NPF_6$ (0.2 mmol). Purification by PTLC (silica gel, hexanes:CH_2Cl_2 = 4:1) afforded **69** (10.7 mg, 27%) as a colorless oil; R_f = 0.31

(hexanes:CH_2Cl_2 = 4:1); ^{1}H NMR (400 MHz, $CDCl_3$) δ 7.13 (d, J = 8.3 Hz, 2H), 6.85 (d, J = 8.3 Hz, 2H), 3.80 (s, 3H), 2.74–2.57 (m, 2H), 1.96–1.84 (m, 2H), 1.41 (d, J = 21.4 Hz, 6H); ^{13}C NMR (126 MHz, $CDCl_3$) δ 157.9, 134.2, 129.3, 114.0, 95.5 (d, J_{C-F} = 165.5 Hz), 55.4, 43.7 (d, J_{C-F} = 22.8 Hz), 29.50 (d, J_{C-F} = 5.5 Hz), 26.8 (d, J_{C-F} = 24.9 Hz); ^{19}F NMR (376 MHz, $CDCl_3$): δ (−140.8) – (−141.3) (m); HRMS (ESI) m/z: $[M+Na]^+$ calcd for $C_{12}H_{17}FONa$ 219.1156; found 219.1155.

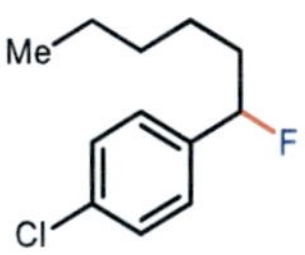

1-Chloro-4-(1-fluorohexyl)benzene (70)

General procedure was applied with alkyl trifluoroborate **S26b** (0.2 mmol), potassium hydrogen difluoride (0.4 mmol), 1,4,7,10,13,16-hexaoxacyclooctadecane (0.4 mmol), and ^{n}Bu_4NPF_6 (0.2 mmol). The reaction mixture was electrolyzed at a constant current of 5.0 mA for 2.0 h. Purification by PTLC (silica gel, 100% hexanes) afforded **70** (18.9 mg, 44%) as a colorless oil; R_f = 0.58 (hexanes); ^{1}H NMR (500 MHz, $CDCl_3$): δ 7.34 (d, J = 8.3 Hz, 2H), 7.25 (d, J = 8.2 Hz, 2H), 5.39 (ddd, J = 47.7, 7.9, 5.0 Hz, 1H), 2.01–1.86 (m, 1H), 1.86–1.69 (m, 1H), 1.50–1.40 (m, 1H), 1.39–1.24 (m, 5H), 0.88 (t, J = 6.9 Hz, 3H); ^{13}C NMR (126 MHz, $CDCl_3$): δ 139.3 (d, J_{C-F} = 20.2 Hz), 134.0 (d, J_{C-F} = 2.3 Hz), 128.7, 127.1 (d, J_{C-F} = 6.8 Hz), 94.1 (d, J_{C-F} = 171.0 Hz), 37.3 (d, J_{C-F} = 23.4 Hz), 31.6, 24.8 (d, J_{C-F} = 4.3 Hz), 22.6, 14.1; ^{19}F NMR (376 MHz, $CDCl_3$): δ (−174.0) – (−174.3) (m); HRMS (EI) m/z: $[M]^+$ calcd for $C_{12}H_{16}ClF$ 214.0925; found 214.0926.

1-Chloroadamantane (71)

General procedure was applied with alkyl trifluoroborate **S16b** (0.2 mmol), potassium chloride (0.4 mmol), 1,4,7,10,13,16-hexaoxacyclooctadecane (0.4 mmol), and ^{n}Bu_4NClO_4 (0.2 mmol). Purification by PTLC (silica gel, 100% hexanes) afforded **71** (20.2 mg, 59%) as a white solid; R_f = 0.69 (hexanes). Spectroscopic data matches with previously reported data [42].

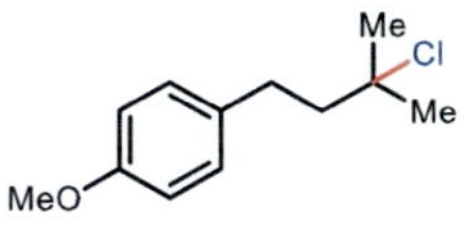

1-(3-Chloro-3-methylbutyl)-4-methoxybenzene (72)

General procedure was applied with alkyl trifluoroborate **S23b** (0.2 mmol), potassium chloride (0.4 mmol), 1,4,7,10,13,16-hexaoxacyclooctadecane (0.4 mmol), and $^nBu_4NClO_4$ (0.2 mmol). Purification by PTLC (silica gel, hexanes:CH_2Cl_2 = 4:1) afforded **72** (8.5 mg, 20%) as a colorless oil; R_f = 0.33 (hexanes:CH_2Cl_2 = 4:1). Spectroscopic data matches with previously reported data [27].

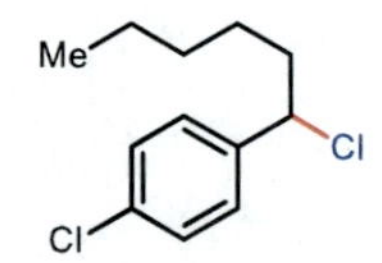

1-Chloro-4-(1-chlorohexyl)benzene (73)

General procedure was applied with alkyl trifluoroborate **S26b** (0.2 mmol), potassium chloride (0.4 mmol), 1,4,7,10,13,16-hexaoxacyclooctadecane (0.4 mmol), and $^nBu_4NClO_4$ (0.2 mmol). The reaction mixture was electrolyzed at a constant current of 5.0 mA for 2.0 h. Purification by PTLC (silica gel, 100% hexanes) afforded **73** (20.8 mg, 45%) as a colorless oil; R_f = 0.69 (hexanes); ^1H NMR (500 MHz, $CDCl_3$): δ 7.35–7.28 (m, 4H), 4.81 (dd, J = 8.0, 6.6 Hz, 1H), 2.15–1.92 (m, 2H), 1.51–1.39 (m, 1H), 1.35–1.22 (m, 5H), 0.92–0.84 (m, 3H); ^{13}C NMR (126 MHz, $CDCl_3$): δ 140.7, 134.0, 128.9, 128.5, 63.0, 40.1, 31.3, 26.8, 22.6, 14.10; HRMS (EI) *m*/*z*: $[M]^+$ calcd for $C_{12}H_{16}Cl_2$ 230.0629; found 230.0629.

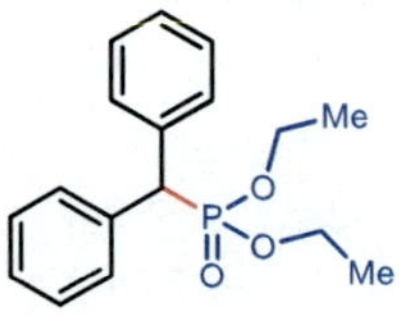

Diethyl benzhydrylphosphonate (74)

General procedure was applied with alkyl trifluoroborate **S4b** (0.2 mmol), triethyl phosphite (0.6 mmol), and $^nBu_4NClO_4$ (0.2 mmol). Purification by flash column chromatography (silica gel, hexanes:acetone = 8:1) afforded **74** (28.8 mg, 47%) as a colorless oil; R_f = 0.5 (hexanes:acetone = 2:1); ^1H NMR (400 MHz, $CDCl_3$): δ 7.58–7.49 (m, 4H), 7.36–7.27 (m, 4H), 7.26–7.18 (m, 2H), 4.46–4.40 (m, 1H), 4.08–3.71 (m, 4H), 1.11 (t, *J* = 7.2 Hz, 6H); ^{13}C NMR (101 MHz, $CDCl_3$): δ 137.0 (d, $J_{C\text{–}P}$ = 5.3 Hz), 129.6 (d, $J_{C\text{–}P}$ = 7.9 Hz), 128.7 (d, $J_{C\text{–}P}$ = 1.3 Hz), 127.3 (d, $J_{C\text{–}P}$ = 2.3 Hz), 62.8 (d, $J_{C\text{–}P}$ = 7.6 Hz), 51.5 (d, $J_{C\text{–}P}$ = 137.0 Hz), 16.4 (d, $J_{C\text{–}P}$ = 5.9 Hz); ^{31}P NMR (243 MHz, $CDCl_3$) δ 24.5; HRMS (ESI) *m*/*z*: $[M+Na]^+$ calcd for $C_{17}H_{21}O_3PNa$ 327.1126; found 327.1141.

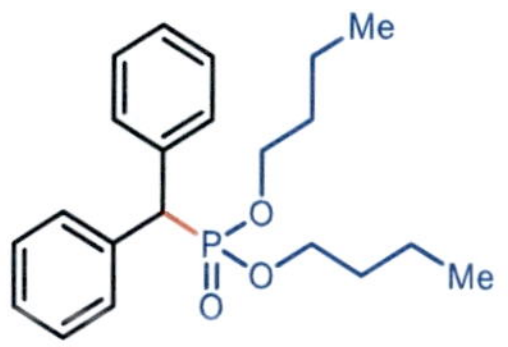

Dibutyl benzhydrylphosphonate (75)
General procedure was applied with alkyl trifluoroborate **S4b** (0.2 mmol), tributyl phosphite (0.6 mmol), and $^{n}Bu_4NClO_4$ (0.2 mmol). Purification by flash column chromatography (silica gel, hexanes:EtOAc = 15:1) afforded **75** (37.8 mg, 52%) as a white solid; m.p. 56–58 °C; R_f = 4:1 (hexanes:EtOAc = 4:1); ^{1}H NMR (400 MHz, $CDCl_3$): δ 7.55–7.50 (m, 4H), 7.31 (dd, J = 8.4, 6.7 Hz, 4H), 7.23 (td, J = 7.3, 1.4 Hz, 2H), 4.43 (d, J = 25.0 Hz, 1H), 3.96–3.70 (m, 4H), 1.49–1.37 (m, 4H), 1.23–1.16 (m, 4H), 0.81 (t, J = 7.4 Hz, 6H); ^{13}C NMR (101 MHz, $CDCl_3$): δ 137.0 (d, J_{C-P} = 5.1 Hz), 129.6 (d, J_{C-P} = 8.4 Hz), 128.7 (d, J_{C-P} = 1.4 Hz), 127.2 (d, J_{C-P} = 1.9 Hz), 66.4 (d, J_{C-P} = 7.3 Hz), 51.4 (d, J_{C-P} = 138.0 Hz), 32.5 (d, J_{C-P} = 5.9 Hz), 18.7, 13.7; ^{31}P NMR (243 MHz, $CDCl_3$): δ 24.4; HRMS (ESI) m/z: $[M+H]^+$ calcd for $C_{21}H_{30}O_3P$ 361.1933; found 361.1928.

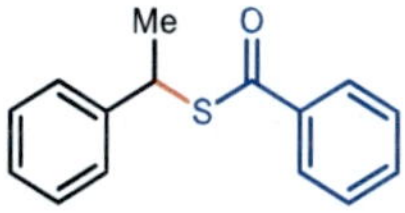

***S*-(1-Phenylethyl) benzothioate (76)**
General procedure was applied with alkyl trifluoroborate **S1b** (0.2 mmol), benzothioic *S*-acid (0.6 mmol), and $^{n}Bu_4NClO_4$ (0.2 mmol). Purification by flash column chromatography (silica gel, hexanes:EtOAc = 40:1) afforded **76** (25.6 mg, 53%) as a colorless oil; R_f = 0.7 (hexanes:EtOAc = 10:1); ^{1}H NMR (400 MHz): δ 7.95 (d, J = 6.9 Hz, 2H), 7.56 (t, J = 7.4 Hz, 1H), 7.44 (t, J = 7.7 Hz, 4H), 7.35 (t, J = 7.7 Hz, 2H), 7.30–7.25 (m, 1H), 4.97 (q, J = 7.2 Hz, 1H), 1.78 (d, J = 7.2 Hz, 3H); ^{13}C NMR (101 MHz, $CDCl_3$): δ 191.3, 142.8, 137.1, 133.5, 128.8, 128.7, 127.5, 127.4, 43.2, 22.5; HRMS (ESI) m/z: $[M+Na]^+$ calcd for $C_{15}H_{14}OSNa$ 265.0663; found 265.0660.

Dodecyl(1-phenylethyl)sulfane (77)

General procedure was applied with alkyl trifluoroborate **S1b** (0.2 mmol), dodecane-1-thiol (0.6 mmol), and $^{n}Bu_4NClO_4$ (0.2 mmol). The reaction mixture was electrolyzed at a constant current of 5.0 mA for 1.5 h. Purification by flash column chromatography (silica gel, hexanes:CH_2Cl_2 = 15:1) afforded **77** (26.4 mg, 43%) as a colorless oil; R_f = 0.59 (hexanes:CH_2Cl_2 = 4:1); ^1H NMR (400 MHz, $CDCl_3$): δ 7.40–7.19 (m, 5H), 3.95 (q, J = 7.0 Hz, 1H), 2.40–2.20 (m, 2H), 1.66–1.12 (m, 23H), 0.89 (t, J = 6.6 Hz, 3H); ^{13}C NMR (101 MHz, $CDCl_3$): δ 144.4, 128.6, 127.4, 127.1, 44.2, 32.1, 31.5, 29.8, 29.8, 29.7, 29.6, 29.5, 29.3, 29.1, 22.8, 22.8, 14.3; HRMS (ESI) m/z: $[M+Na]^+$ calcd for $C_{20}H_{34}SNa$ 329.2279; found 329.2277.

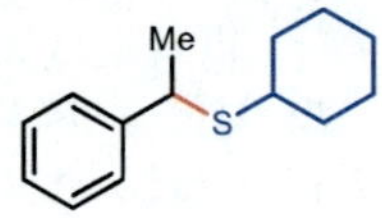

Cyclohexyl(1-phenylethyl)sulfane (78)

General procedure was applied with alkyl trifluoroborate **S1b** (0.2 mmol), cyclohexanethiol (0.6 mmol), and $^{n}Bu_4NClO_4$ (0.2 mmol). The reaction mixture was electrolyzed at a constant current of 5.0 mA for 1.5 h. Purification by flash column chromatography (silica gel, hexanes:CH_2Cl_2 = 4:1) afforded **78** (20.5 mg, 47%) as a colorless oil; R_f = 0.42 (hexanes:CH_2Cl_2 = 4:1). Spectroscopic data matches with previously reported data [43].

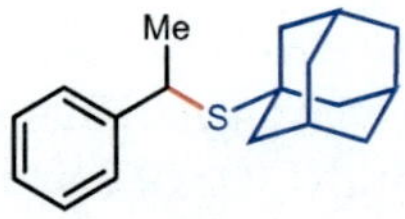

(Adamantan-1-yl)(1-phenylethyl)sulfane (79)

General procedure was applied with alkyl trifluoroborate **S1b** (0.2 mmol), adamantane-1-thiol (0.6 mmol), and $^{n}Bu_4NClO_4$ (0.2 mmol). The reaction mixture was electrolyzed at a constant current of 5.0 mA for 1.5 h. Purification by flash column chromatography (silica gel, hexanes:CH_2Cl_2 = 24:1) afforded **79** (21.9 mg, 40%) as a colorless oil; R_f = 0.28 (hexanes:CH_2Cl_2 = 24:1). Spectroscopic data matches with previously reported data [44].

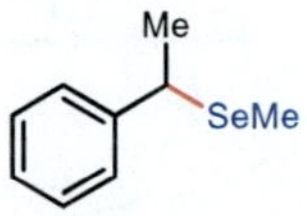

Methyl(1-phenylethyl)selane (80)

General procedure was applied with alkyl trifluoroborate **S1b** (0.2 mmol), 1,2-dimethyldiselane (0.6 mmol), and $^{n}Bu_4NClO_4$ (0.2 mmol). The reaction mixture was electrolyzed at a constant current of 5.0 mA for 1.5 h. Purification by flash column chromatography (silica gel, hexanes:CH_2Cl_2 = 7:3) afforded **80** (17.8 mg, 45%) as a colorless oil; R_f = 0.4 (hexanes:CH_2Cl_2 = 4:1); ^1H NMR (400 MHz, $CDCl_3$): δ 7.36–7.27 (m, 4H), 7.25–7.18 (m, 1H), 4.11 (q, J = 7.1 Hz, 1H), 1.84 (s, 3H), 1.73 (d, J = 7.1 Hz, 3H); ^{13}C NMR (101 MHz, $CDCl_3$): δ 144.4, 128.5, 127.3, 126.9, 37.9, 22.2, 4.55; HRMS (EI) m/z: $[M]^+$ calcd for $C_9H_{12}Se$ 200.0104; found 200.0105. Note: for **71**, the solvent was removed under reduced pressure (120 mbar) at 30 °C and the use of a high vacuum pump was avoided during the whole process due to the volatility of the product.

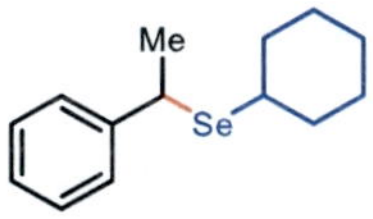

Cyclohexyl(1-phenylethyl)selane (81)

General procedure was applied with alkyl trifluoroborate **S1b** (0.2 mmol), 1,2-dicyclohexyldiselane (0.6 mmol), and $^{n}Bu_4NClO_4$ (0.2 mmol). The reaction mixture was electrolyzed at a constant current of 5.0 mA for 1.5 h. Purification by flash column chromatography (silica gel, hexanes:CH_2Cl_2 = 4:1) afforded **81** (22.0 mg, 41%) as a colorless oil; R_f = 0.36 (hexanes:CH_2Cl_2 = 4:1); ^1H NMR (500 MHz, $CDCl_3$): δ 7.34 (d, J = 7.3 Hz, 2H), 7.29 (t, J = 7.5 Hz, 2H), 7.20 (t, J = 7.2 Hz, 1H), 4.23 (q, J = 7.0 Hz, 1H), 2.77–2.66 (m, 1H), 2.04–1.18 (m, 13H); ^{13}C NMR (126 MHz, $CDCl_3$): δ 145.2, 128.5, 127.3, 126.8, 39.8, 36.6, 35.0, 34.5, 26.9, 26.0, 23.3; HRMS (ESI) m/z: $[M+Na]^+$ calcd for $C_{14}H_{20}SeNa$ 291.0628; found 291.0621.

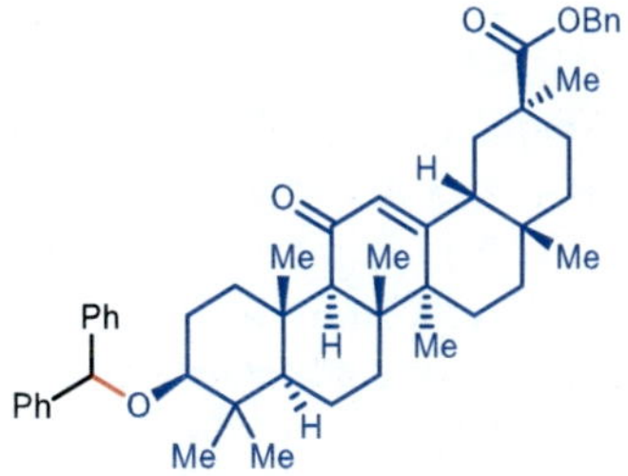

Benzyl (2*S*,4a*S*,6a*S*,6b*R*,8a*R*,10*S*,12a*S*,12b*R*,14b*R*)-10-(Benzhydryloxy)-2,4a,6a,6b,9,9,12a-heptamethyl-13-oxo-1,2,3,4,4a,5,6,6a,6b,7,8,8a,9,10,11,12,12a,12b,13,14b-icosahydropicene-2-carboxylate (82)

General procedure was applied with alkyl trifluoroborate **S4b** (0.2 mmol), benzyl (2*S*,4a*S*,6a*S*,6b*R*,8a*R*,10*S*,12a*S*,12b*R*,14b*R*)-10-hydroxy-2,4a,6a,6b,9,9,12a-heptamethyl-13-oxo-1,2,3,4,4a,5,6,6a,6b,7,8,8a,9,10,11,12,12a,12b,13,14b-icosahydropicene-2-carboxylate (0.4 mmol), and $^{n}Bu_4NClO_4$ (0.2 mmol). The reaction mixture was electrolyzed at a constant current of 10.0 mA for 2.0 h. Purification by flash column chromatography (silica gel, hexanes:EtOAc = 30:1) afforded **82** (48.1 mg, 33%) as a colorless oil; R_f = 0.4 (hexanes:EtOAc = 10:1); ^{1}H NMR (400 MHz, $CDCl_3$): δ 7.44–7.15 (m, 15H), 5.51 (d, *J* = 5.4 Hz, 2H), 5.27–5.02 (m, 2H), 2.96–2.92 (m, 1H), 2.75–2.71 (m, 1H), 2.23 (s, 1H), 2.04–1.66 (m, 6H), 1.59–0.86 (m, 30H), 0.71 (s, 3H), 0.61–0.58 (m, 1H); ^{13}C NMR (101 MHz, $CDCl_3$): δ 200.2, 176.2, 168.9, 144.2, 142.7, 136.1, 128.6, 128.5, 128.3, 128.2, 128.1, 128.0, 127.9, 127.3, 126.8, 126.6, 126.5, 83.1, 80.1, 66.2, 61.8, 55.2, 48.2, 45.4, 44.0, 43.1, 41.0, 39.2, 38.9, 37.6, 37.0, 32.7, 31.8, 31.6, 31.2, 28.4, 28.3, 28.2, 26.4, 26.4, 23.3, 22.7, 22.5, 18.7, 17.5, 16.9, 16.4, 14.1; HRMS (ESI) *m*/*z*: $[M+H]^+$ calcd for $C_{50}H_{63}O_4$ 727.4726; found 727.4724.

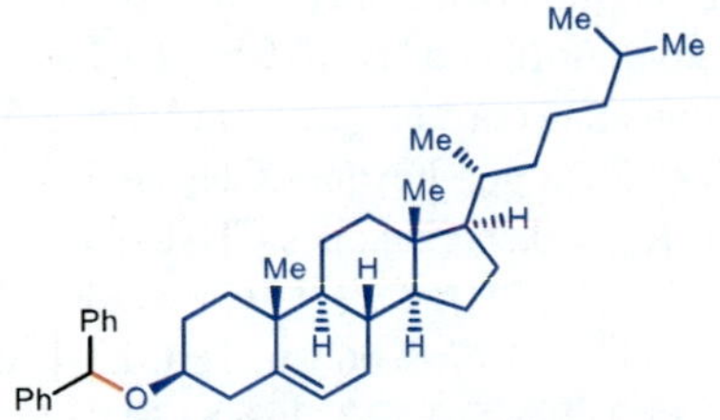

(3*S*,8*S*,9*S*,10*R*,13*R*,14*S*,17*R*)-3-(Benzhydryloxy)-10,13-dimethyl-17-((*R*)-6-methylheptan-2-yl)-2,3,4,7,8,9,10,11,12,13,14,15,16,17-tetradecahydro-1*H*-cyclopenta[*a*]phenanthrene (83)

General procedure was applied with alkyl trifluoroborate **S4b** (0.2 mmol), (3*S*,8*S*,9*S*,10*R*,13*R*,14*S*,17*R*)-10,13-dimethyl-17-((*R*)-6-methylheptan-2-yl)-2,3,4,7,8,9,10,11,12,13,14,15,16,17-tetradecahydro-1*H*-cyclopenta[*a*]phenanthren-3-ol (0.4 mmol), and $^{n}Bu_4NClO_4$ (0.2 mmol). The reaction mixture was electrolyzed at a constant current of 10.0 mA for 2.0 h. Purification by flash column chromatography (silica gel, hexanes:CH_2Cl_2 = 8:1) afforded **83** (45.5 mg, 41%) as a white solid; m.p. 121–123 °C; R_f = 0.7 (hexanes:CH_2Cl_2 = 2:1); ^{1}H NMR (400 MHz, $CDCl_3$): δ 7.38–7.17 (m, 10H), 5.55 (s, 1H), 5.29 (s, 1H), 3.30–3.25 (m, 1H), 2.37 (t, *J* = 8.4 Hz, 2H), 2.02–1.88 (m, 3H), 1.88–1.74 (m, 2H), 1.67–0.78 (m, 33H), 0.65 (s, 3H); ^{13}C NMR (101 MHz, $CDCl_3$): δ 143.1, 141.2, 128.5, 127.4, 127.2, 121.7, 80.5, 77.0, 56.9, 56.3, 50.3, 42.5, 39.9, 39.7, 39.5, 37.4, 37.0, 36.3, 35.9,

32.1, 32.0, 28.8, 28.4, 28.2, 24.4, 24.0, 23.0, 22.7, 21.2, 19.6, 18.9, 12.0; HRMS (ESI) *m*/*z*: $[M+Na]^+$ calcd for $C_{40}H_{56}ONa$ 575.4228; found 575.4226.

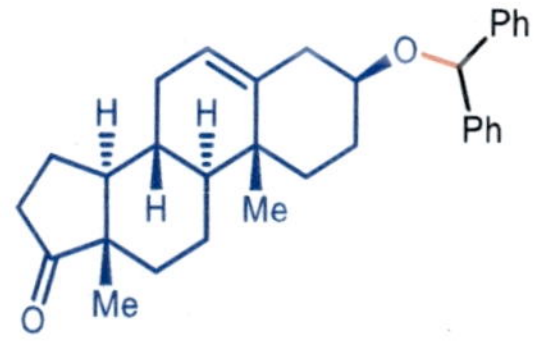

(3*S*,8*R*,9*S*,10*R*,13*S*,14*S*)-3-(Benzhydryloxy)-10,13-dimethyl-1,2,3,4,7,8,9,10,11,12,13,14,15,16-tetradecahydro-17*H*-cyclopenta[*a*]phenanthren-17-one (84)

General procedure was applied with alkyl trifluoroborate **S4b** (0.2 mmol), (3*S*,8*R*,9*S*,10*R*,13*S*,14*S*)-3-hydroxy-10,13-dimethyl-1,2,3,4,7,8,9,10,11,12,13,14,15,16-tetradecahydro-17H-cyclopenta[*a*]phenanthren-17-one (0.4 mmol), and $^nBu_4NClO_4$ (0.2 mmol). The reaction mixture was electrolyzed at a constant current of 10.0 mA for 2.0 h. Purification by flash column chromatography (silica gel, hexanes:CH_2Cl_2 = 20:1) afforded **84** (45.7 mg, 50%) as a white solid; m.p. 143–145 °C; R_f = 0.3 (hexanes:CH_2Cl_2 = 10:1); ^{1}H NMR (400 MHz, $CDCl_3$): δ 7.43–7.17 (m, 10H), 5.55 (s, 1H), 5.32 (d, *J* = 5.3 Hz, 1H), 3.35–3.23 (m, 1H), 2.47–2.36 (m, 3H), 2.14–1.77 (m, 6H), 1.73–1.37 (m, 6H), 1.32–1.16 (m, 2H), 1.06–0.82 (m, 8H); ^{13}C NMR (101 MHz, $CDCl_3$): δ 221.3, 143.0, 143.0, 141.4, 128.6, 128.5, 128.5, 127.4, 127.2, 126.7, 120.9, 80.5, 76.8, 51.9, 50.4, 47.7, 39.4, 37.3, 37.1, 36.0, 31.6, 31.6, 30.9, 28.7, 22.0, 20.5, 19.6, 13.7; HRMS (ESI) *m*/*z*: $[M+Na]^+$ calcd for $C_{32}H_{38}O_2Na$ 477.2770; found 477.2767.

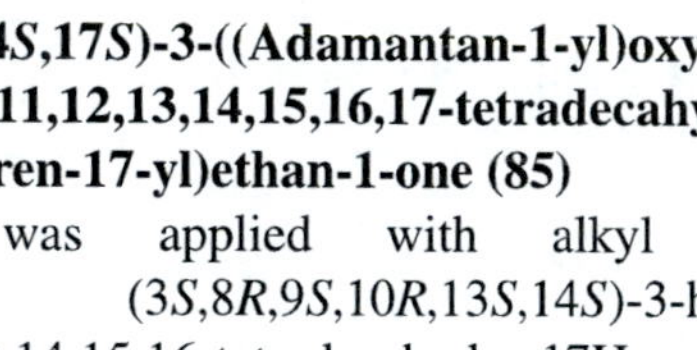

1-((3*S*,8*S*,9*S*,10*R*,13*S*,14*S*,17*S*)-3-((Adamantan-1-yl)oxy)-10,13-dimethyl-2,3,4,7,8,9,10,11,12,13,14,15,16,17-tetradecahydro-1*H*-cyclopenta[*a*]phenanthren-17-yl)ethan-1-one (85)

General procedure was applied with alkyl trifluoroborate **S16b** (0.2 mmol), (3*S*,8*R*,9*S*,10*R*,13*S*,14*S*)-3-hydroxy-10,13-dimethyl-1,2,3,4,7,8,9,10,11,12,13,14,15,16-tetradecahydro-17H-cyclopenta[*a*]phenanthren-17-one (0.4 mmol), and $^nBu_4NClO_4$ (0.2 mmol). The reaction mixture was electrolyzed at a constant current of 10.0 mA for 2.0 h. Purification by flash column chromatography (silica gel, hexanes:EtOAc = 15:1) afforded **85** (32.6 mg, 36%)

as a white solid; m.p. 197–199 °C; R_f = 0.6 (hexanes:EtOAc = 5:1); ^{1}H NMR (400 MHz, $CDCl_3$): δ 5.30 (d, *J* = 5.5 Hz, 1H), 3.50–3.42 (m, 1H), 2.55–2.51 (m, 1H), 2.32–2.26 (m, 1H), 2.20–1.07 (m, 35H), 0.98 (s, 4H), 0.62 (s, 3H); ^{13}C NMR (101 MHz, $CDCl_3$): δ 209.8, 142.2, 120.8, 72.8, 69.4, 63.9, 57.1, 50.3, 44.2, 42.7, 42.4, 39.0, 37.9, 36.8, 36.6, 32.0, 32.0, 31.7, 31.7, 30.7, 24.6, 22.9, 21.2, 19.5, 13.4; HRMS (ESI) *m*/*z*: $[M+Na]^+$ calcd for $C_{31}H_{46}O_2Na$ 473.3396; found 473.3394.

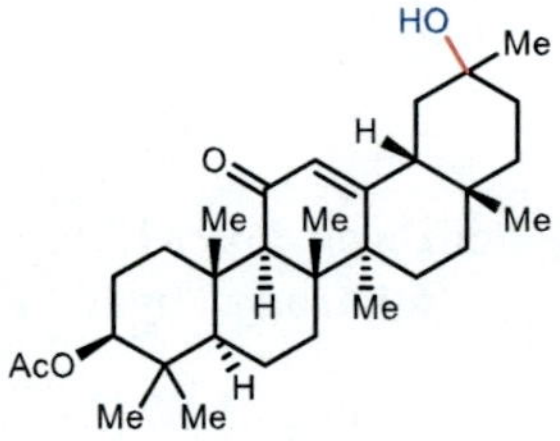

(3*S*,4a*R*,6a*R*,6b*S*,8a*S*,12a*R*,14a*R*,14b*S*)-11-Hydroxy-4,4,6a,6b,8a,11,14b-heptamethyl-14-oxo-1,2,3,4,4a,5,6,6a,6b,7,8,8a,9,10,11,12,12a,14,14a,14b-icosahydropicen-3-yl acetate (86)

General procedure was applied with alkyl trifluoroborate **S8b** (0.2 mmol), H_2O (0.1 mL), and ^{n}Bu$_4$NPF$_6$ (0.2 mmol). Purification by flash column chromatography (silica gel, hexanes:EtOAc = 3:1) afforded **86** (85.1 mg, 88%, the ratio of the isomer **A** and **B** of **86** to be 1:1.1) as a white solid. Isomer **A**: m.p. 214–216 °C; R_f = 0.45 (hexanes:EtOAAc = 3:2); ^{1}H NMR (400 MHz, $CDCl_3$): δ 5.63 (s, 1H), 4.51 (dd, *J* = 11.6, 4.6 Hz, 1H), 2.84–2.71 (m, 1H), 2.42–2.30 (m, 2H), 2.04 (s, 3H), 2.10–1.94 (m, 1H), 1.89–1.76 (m, 2H), 1.76–0.74 (m, 16H), 1.34 (s, 3H) 1.22 (s, 3H), 1.15 (s, 3H), 1.13 (s, 3H), 0.87 (s, 9H); ^{13}C NMR (101 MHz, $CDCl_3$): δ 200.2, 171.2, 169.9, 128.3, 80.8, 69.5, 61.8, 55.1, 46.7, 45.6, 44.4, 43.4, 38.9, 38.2, 37.1, 35.6, 34.1, 32.8, 32.0, 31.7, 28.4, 28.2, 26.6, 26.1, 23.7, 23.6, 21.5, 18.8, 17.5, 16.8, 16.5; HRMS (ESI) *m*/*z*: $[M+H]^+$ Calcd for $C_{31}H_{49}O_4$ 485.3631; found 485.3629. Isomer **B**: m.p. 263–265 °C; R_f = 0.36 (hexanes:EtOAc = 3:2); ^{1}H NMR (500 MHz, $CDCl_3$): δ 5.59 (s, 1H), 4.50 (dd, *J* = 11.6, 4.5 Hz, 1H), 2.78 (d, *J* = 13.7 Hz, 1H), 2.40–2.30 (m, 1H), 2.20–1.90 (m, 3H), 2.04 (s, 3H), 1.88–1.76 (m, 1H), 1.74–0.75 (m, 16H), 1.37 (s, 3H), 1.24 (s, 3H), 1.15 (s, 3H), 1.12 (s, 3H), 0.87 (s, 6H), 0.86 (s, 3H); ^{13}C NMR (101 MHz, $CDCl_3$): δ 200.3, 171.1, 168.9, 128.4, 80.7, 71.5, 61.8, 55.1, 49.6, 45.8, 45.6, 43.4, 38.9, 38.4, 38.2, 37.1, 35.6, 32.8, 32.6, 28.3, 28.2, 26.5, 26.5, 25.3, 23.7, 23.6, 21.5, 18.9, 17.5, 16.8, 16.6; HRMS (ESI) *m*/*z*: $[M+H]^+$ Calcd for $C_{31}H_{49}O_4$ 485.3631; found 485.3628.

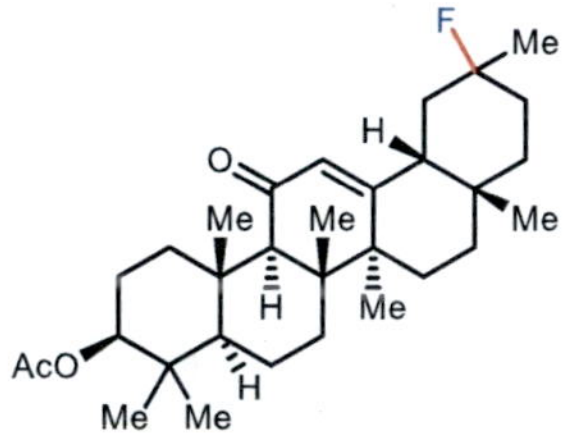

(3*S*,4a*R*,6a*R*,6b*S*,8a*S*,12a*R*,14a*R*,14b*S*)-11-Fluoro-4,4,6a,6b,8a,11,14b-heptamethyl-14-oxo-1,2,3,4,4a,5,6,6a,6b,7,8,8a,9,10,11,12,12a,14,14a,14b-icosahydropicen-3-yl acetate (87)

General procedure was applied with alkyl trifluoroborate **S8b** (0.2 mmol), potassium hydrogen difluoride (0.4 mmol), 1,4,7,10,13,16-hexaoxacyclooctadecane (0.4 mmol), and $^{n}Bu_4NPF_6$ (0.2 mmol). The reaction mixture was electrolyzed at a constant current of 5.0 mA for 2.0 h. Purification by flash column chromatography (silica gel, hexanes:EtOAc = 19:1) afforded **87** (44.9 mg, 46%, the ratio of the isomers of **87** to be 1.9:1) as a white solid; m.p. 261–263 °C; $R_f = 0.5$ (hexanes:EtOAc = 4:1); ^{1}H NMR (500 MHz, $CDCl_3$, for both diastereomers) (the integration at 5.63 ppm and 5.59 ppm indicated the ratio of the two isomers of **87** to be 1.9:1): 5.63 (s, 0.66H), 5.59 (s, 0.34H), 4.51 (dd, J = 11.7, 4.5 Hz, 1H), 2.85–2.72 (m, 1H), 2.41–2.30 (m, 2H), 2.24–1.90 (m, 3H), 2.05 (s, 3H), 1.9–0.77 (m, 36H); ^{13}C NMR (126 MHz, $CDCl_3$): δ 200.1, 200.1, 171.1, 171.1, 168.8, 167.8, 128.5, 128.5, 96.1 (d, $J_{C\text{–}F}$ = 168.5 Hz), 93.7 (d, $J_{C\text{–}F}$ = 169.2 Hz), 80.7, 80.7, 61.9, 61.8, 55.1, 55.1, 49.4, 49.3, 46.4, 45.5, 45.5, 43.4, 43.3, 43.0, 42.8 42.1, 41.9, 38.9, 38.9, 38.2, 38.2, 38.2, 37.1, 35.3, 33.2, 33.1, 32.8, 32.7, 32.7, 32.3, 32.1, 31.7, 28.2, 28.2, 28.0, 28.0, 27.9, 27.7, 26.5, 26.5, 26.4, 26.1, 23.7, 23.7, 23.6, 23.4, 23.3, 23.1, 21.4, 18.8, 18.8, 17.5, 16.8, 16.5, 16.5 (observed complexity is due to C–F coupling); ^{19}F NMR (376 MHz, $CDCl_3$) (the integration at (−124.1) – (−124.4) ppm and (−154.4) – (−154.6) ppm indicated the ratio of the isomers of **75** to be 1:1.9): δ (−124.1) – (−124.4) (m, 1F), (−154.4) – (−154.6) (m, 1.9F); HRMS (ESI) *m*/*z*: $[M+H]^+$ Calcd for $C_{31}H_{48}FO_3$ 487.3588; found 487.3586.

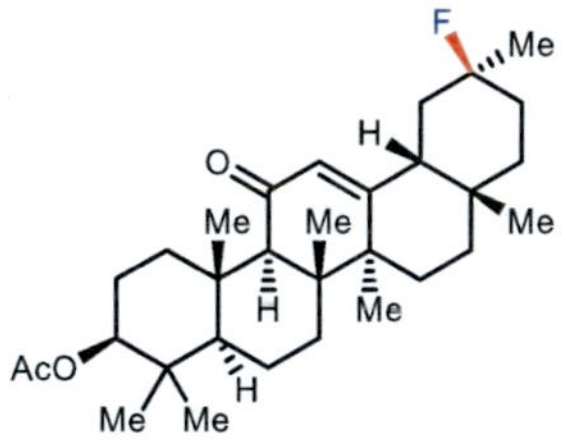

(3*S*,4a*R*,6a*R*,6b*S*,8a*S*,11*S*,12a*R*,14a*R*,14b*S*)-11-Fluoro-4,4,6a,6b,8a,11,14b-heptamethyl-14-oxo-1,2,3,4,4a,5,6,6a,6b,7,8,8a,9,10,11,12,12a,14,14a,14b-icosahydropicen-3-yl acetate ((11*S*)-87)

The vapor diffusion crystallization method using CH_2Cl_2/pentane afforded **(11*S*)-87** as a colorless crystal; m.p. 263–265 °C; $R_f = 0.5$ (hexanes:EtOAc = 4:1); ^{1}H NMR (500 MHz, $CDCl_3$): δ 5.63 (s, 1H), 4.51 (dd, J = 11.7, 4.4 Hz, 1H), 2.79 (d, J = 13.6 Hz, 1H), 2.41–2.27 (m, 2H), 2.05 (s, 3H), 2.02–1.90 (m, 1H), 1.90–1.76 (m, 2H), 1.77–0.73 (m, 15H), 1.34 (s, 6H), 1.16 (s, 3H), 1.14 (s, 3H), 0.87 (s, 9H); ^{13}C NMR (126 MHz, $CDCl_3$): δ 200.1, 171.1, 168.7, 128.6, 93.7 (d, $J_{C\text{–}F}$ = 169.2 Hz), 80.8, 61.9, 55.2, 46.4, 45.6, 43.4, 42.2, 42.0, 39.0, 38.2, 37.1, 35.4, 32.8, 32.3, 32.1, 31.8, 28.2, 27.9, 27.7, 26.6, 26.1, 23.7, 23.7, 21.5, 18.9, 17.5, 16.8, 16.5 (observed complexity is due to C–F coupling); ^{19}F NMR (376 MHz, $CDCl_3$): (−154.4) – (−154.6) (m); HRMS (ESI) *m*/*z*: $[M+H]^+$ Calcd for $C_{31}H_{48}FO_3$ 487.3588; found 487.3586.

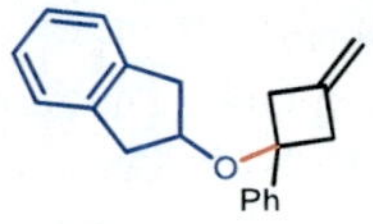

2-(3-Methylene-1-phenylcyclobutoxy)-2,3-dihydro-1*H*-indene (88)

General procedure was applied with alkyl trifluoroborate **S12b** (0.2 mmol), 2,3-dihydro-1*H*-inden-2-ol (0.6 mmol), and ^{n}Bu$_4$NPF$_6$ (0.2 mmol). The reaction mixture was electrolyzed at a constant current of 5.0 mA for 3.0 h. Purification by flash column chromatography (silica gel, hexanes:EtOAc = 30:1) afforded **88** (20.5 mg, 37%) as a colorless oil; $R_f = 0.6$ (hexanes:EtOAc = 10:1); ^{1}H NMR (400 MHz, $CDCl_3$) δ 7.48 (d, J = 8.2 Hz, 2H), 7.42–7.31 (m, 3H), 7.09 (s, 4H), 4.93 (s, 2H), 4.16 (p, J = 7.1 Hz, 1H), 3.17 (s, 4H), 2.84 (d, J = 7.1 Hz, 4H); ^{13}C NMR (100 MHz, $CDCl_3$) δ 143.5, 140.7, 140.7, 128.3, 127.4, 126.6, 126.4, 124.4, 107.5, 77.9, 76.2, 44.5, 40.3; HRMS (ESI) *m*/*z*: $[M+Na]^+$ calcd for $C_{20}H_{20}ONa$ 299.1412; found 299.1414.

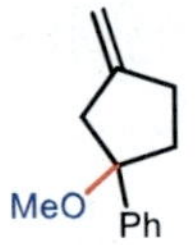

(1-Methoxy-3-methylenecyclopentyl)benzene (89)

General procedure was applied with alkyl trifluoroborate **S13b** (0.2 mmol), methanol (0.1 mL), and ^{n}Bu$_4$NPF$_6$ (0.2 mmol). The reaction mixture was electrolyzed at a constant current of 5.0 mA for 3.0 h. Purification by flash column chromatography (silica gel, hexanes:EtOAc = 30:1) afforded **89** (25.0 mg, 66%) as a colorless oil; $R_f = 0.5$ (hexanes:EtOAc = 10:1); ^{1}H NMR (400 MHz, $CDCl_3$) δ 7.41–7.38 (m, 2H), 7.37–7.32 (m, 2H), 7.31–7.26 (m, 1H), 4.94 (d, J = 16.6 Hz, 2H), 2.97 (s, 3H), 2.87–2.78 (m, 1H), 2.72–2.67 (m, 1H), 2.61–2.54 (m, 1H), 2.47–2.40 (m, 1H), 2.37–2.24 (m, 1H), 2.06–1.96 (m, 1H); ^{13}C NMR (101 MHz, $CDCl_3$) δ 150.0, 128.3,

128.2, 127.3, 126.6, 106.7, 87.0, 50.9, 44.3, 35.8, 29.9; HRMS (ESI) *m/z*: $[M+Na]^+$ calcd for $C_{13}H_{16}ONa$ 211.1099; found 211.1096.

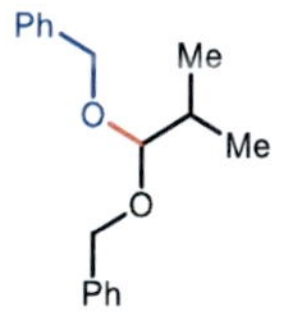

(((2-Methylpropane-1,1-diyl)bis(oxy))bis(methylene))dibenzene (90)
General procedure was applied with alkyl trifluoroborate **S17b** (0.2 mmol), benzyl alcohol (0.6 mmol), and $^nBu_4NClO_4$ (0.2 mmol). The reaction mixture was electrolyzed at a constant current of 5.0 mA for 2.0 h. Purification by flash column chromatography to afford the desired product **90** (5.5 mg, 10%) as a colorless oil; R_f = 0.5 (hexanes:EtOAc = 1:1). Spectroscopic data matches with previously reported data [35].

4.4.12 Unsuitable Trifluoroborate Substrates and Heteroatom Nucleophiles

We tested numerous alkyl trifluoroborates and nucleophiles which are not included in the manuscript.

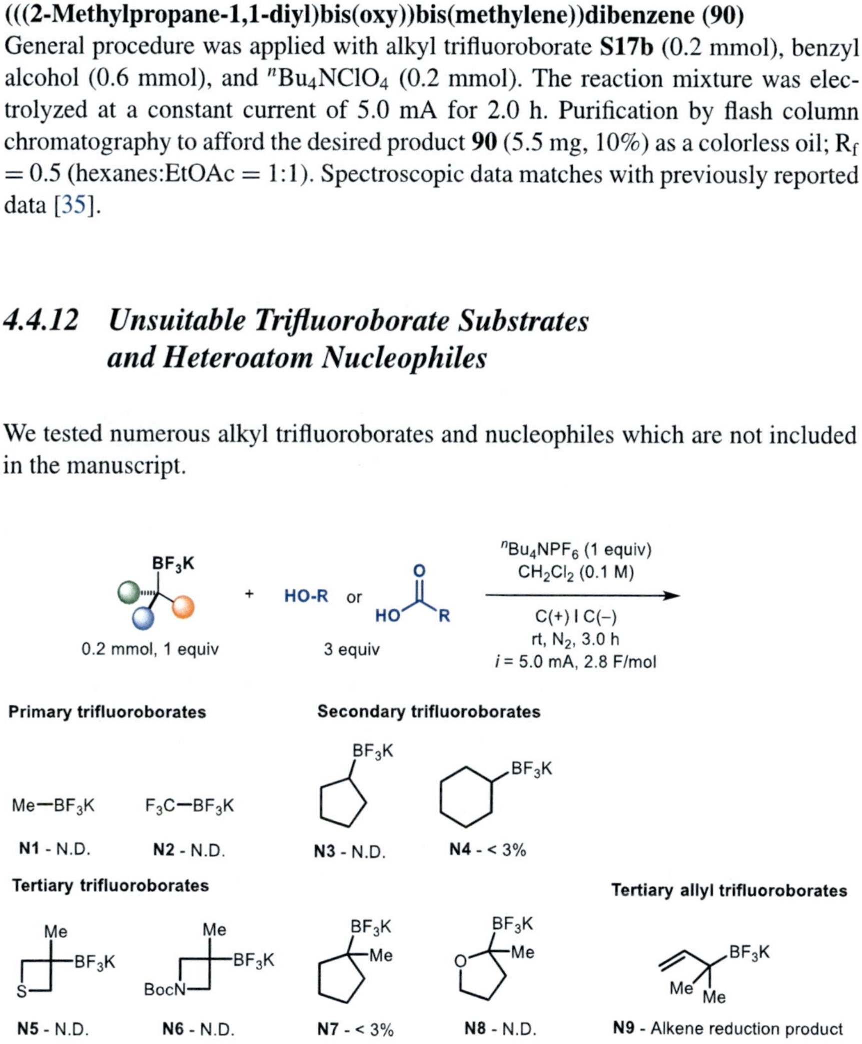

Substrates that generate highly unstable carbocation intermates were generally shown to be unsuitable for the developed reaction. For instance, trifluoroborates based on primary carbon (**N1**, **N2**) or unactivated secondary carbon (**N3**, **N4**) usually provided no or trace amounts of products. Also, the reaction center embedded in a 4- or 5-membered ring did not lead to the desired transformation, presumably due to ring strain of the carbocation intermediate (**N5-N8**). An allylic trifluoroborate substrate successfully underwent the desired C–heteroatom bond formation. However, a product in which the alkene is fully reduced was obtained as a major product (**N9**).

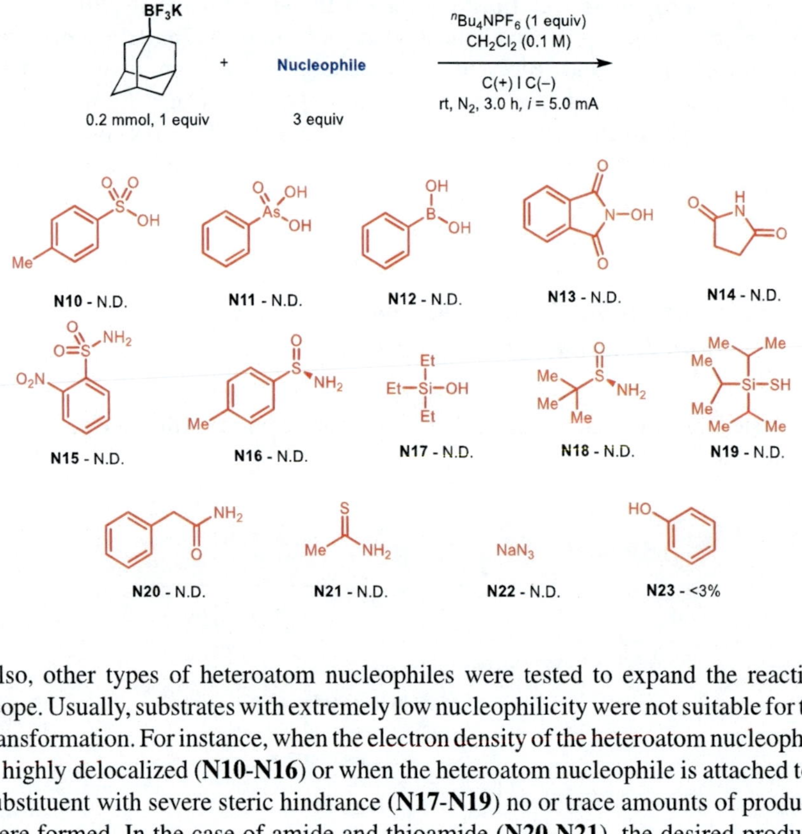

Also, other types of heteroatom nucleophiles were tested to expand the reaction scope. Usually, substrates with extremely low nucleophilicity were not suitable for the transformation. For instance, when the electron density of the heteroatom nucleophile is highly delocalized (**N10-N16**) or when the heteroatom nucleophile is attached to a substituent with severe steric hindrance (**N17-N19**) no or trace amounts of products were formed. In the case of amide and thioamide (**N20-N21**), the desired products were not detected. Also, some of the redox-labile nucleophiles, such as a phenol (**N22**) or an azide (**N23**), were incompatible with the reaction conditions.

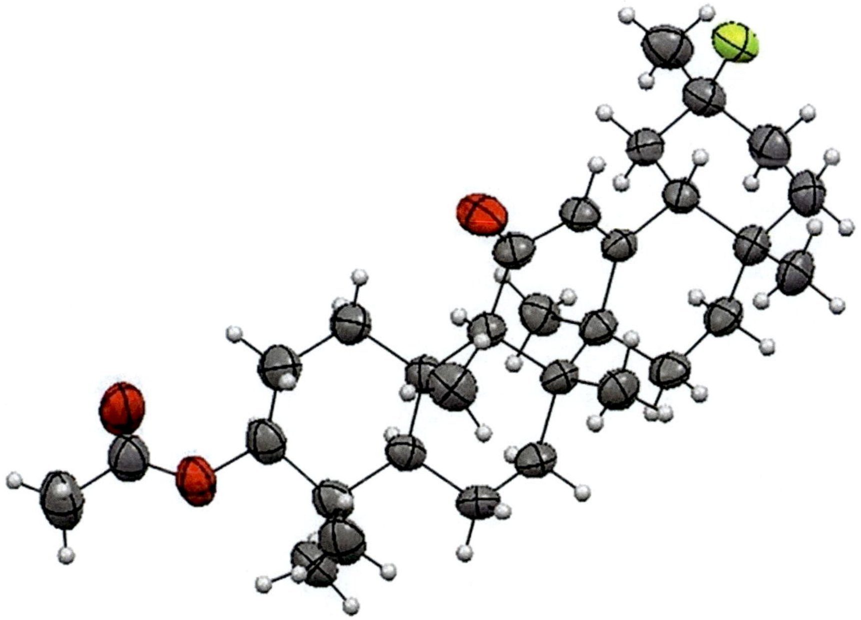

Fig. 4.31 X-ray of compound of **(11*S*)-87**

4.4.13 X-Ray of Compound of (11S)-87

CCDC 2102184

The crystal structure of **(11*S*)-87** was determined by standard crystallographic method. A colorless plate-shaped crystal ($0.04 \times 0.16 \times 0.31$ mm^3) was used for single-crystal X-ray diffraction. The data were collected at 243(2) K using a Bruker D8 Venture equipped with IμS micro-focus sealed tube Mo Kα ($\lambda = 0.71073$ Å) and a PHOTON III M14 detector. Data collection and integration were performed with SMART APEX3 software package (SAINT+) [45]. Absorption correction was performed by multi-scan method implemented in SADABS [46]. The structure was solved by direct methods and refined by full-matrix least-squares on F^2 using SHELXTL program package (version 6.14) [47]. All the non-hydrogen atoms were refined anisotropically, and hydrogen atoms were added to their geometrically ideal positions. A similar structure (different halogen ion, bromide) to compound **(11*S*)-87** has been reported previously (Fig. 4.31, Table 4.12) [48].

Table 4.12 Crystal data and structure refinement for **(11*S*)-87**

Identification code	**(11*S*)-87**
Empirical formula	C31 H47 F O3
Formula weight	486.68
Temperature	253(2) K
Wavelength	1.54178 Å
Crystal system	Monoclinic
Space group	$P2_1$
Unit cell dimensions	a = 14.9111(7) Å α = 90°
	b = 11.4856(6) Å β = 90.103(2)°
	c = 16.4880(8) Å γ = 90°
Volume	2823.8(2) $Å^3$
Z	4
Density (calculated)	1.145 Mg/m^3
Absorption coefficient	0.599 mm^{-1}
F(000)	1064
Crystal size	0.312 × 0.161 × 0.038 mm^3
Theta range for data collection	2.963–77.369°
Index ranges	−18 <= h <= 16, −14 <= k <= 14, −20 <= l <= 20
Reflections collected	43,367
Independent reflections	11,343 [R(int) = 0.0490]
Completeness to theta = 67.679°	96.7%
Absorption correction	Semi-empirical from equivalents
Max. and min. transmission	0.7541 and 0.6441
Refinement method	Full-matrix least-squares on F^2
Data/restraints/parameters	11,343/1/647
Goodness-of-fit on F^2	1.152
Final *R* indices [I > 2sigma(I)]	R1 = 0.0546, wR2 = 0.1254
R indices (all data)	R1 = 0.0621, wR2 = 0.1303
Absolute structure parameter	−0.06(7)
Extinction coefficient	n/a
Largest diff. peak and hole	0.286 and −0.165 e.$Å^{-3}$

References

1. (a) Hartwig JF (2008) Carbon–heteroatom Bond formation catalysed by organometallic complexes. Nature 455:314–322. For examples of important functional molecules with heteroatom substituents: (b) Herzon SB (2017) The mechanism of action of (−)-lomaiviticin A. Acc Chem Res 50:2577–2588. (c) Sather AC, Lee HG, De La Rosa VY, Yang Y, Müller P, Buchwald SL (2015) A Fluorinated ligand enables room-temperature and regioselective Pd-catalyzed fluorination of aryl triflates and bromides. J Am Chem Soc 137:13433–13438. (d) Chen CH, Tang CW (2001) Efficient green organic light-emitting diodes with stericly hindered coumarin dopants. Appl Phys Lett 79:3711–3713. (e) Martin SJ, Bradley DDC, Lane PA, Mellor H, Burn PL (1999) Linear and nonlinear optical properties of the conjugated polymers PPV and MEH-PPV. Phys Rev B 59:15133–15142
2. For review, see: (a) Ruiz-Castillo P, Buchwald SL (2016) Applications of palladium-catalyzed C-N cross-coupling reactions. Chem Rev 116:12564–12649. (b) Chen J-Q, Li J-H, Dong Z-B (2020) A review on the latest progress of Chan-Lam coupling reaction. Adv Synth Catal 362:3311–3331. (c) Barcellos AM, Sacramento M, da Costa GP, Perin G, João Lenardão E, Alves D (2021) Organoboron compounds as versatile reagents in the transition metal-catalyzed C–S, C–Se and C–Te bond formation. Coord Chem Rev 442:214012. (d) Cavedon C, Seeberger PH, Pieber B (2020) Photochemical strategies for carbon–heteroatom bond formation. Eur J Org Chem 2020:1379–1392. (e) Zhu C, Yue H, Jia J, Rueping M (2021) Nickel-catalyzed C-heteroatom cross-coupling reactions under mild conditions via facilitated reductive elimination. Angew Chem Int Ed 60:17810–17831. For a recently reported electrochemical apporoach, see: (f) Walker BR, Manabe S, Brusoe AT, Sevov CS (2021) Mediator-enabled electrocatalysis with ligandless copper for anaerobic Chan-Lam coupling reactions. J Am Chem Soc 143:6257–6265
3. (a) Liu L, Tang Y, Wang K, Huang T, Chen T (2021) Transition-metal-free and base-promoted carbon–heteroatom bond formation via C–N cleavage of benzyl ammonium salts. J Org Chem 86:4159–4170. (b) Fletcher S (2015) The Mitsunobu reaction in the 21st century. Org Chem Front 2:739–752
4. (a) Mao R, Balon J, Hu X (2018) Decarboxylative C(sp^3)–O cross-coupling. Angew Chem Int Ed 57:13624–13628. (b) Mao R, Frey A, Balon J, Hu X (2018) Decarboxylative C(sp^3)–N cross-coupling via synergetic photoredox and copper catalysis. Nat Catal 1:120–126. (c) Liang Y, Zhang X, MacMillan DWC (2018) Decarboxylative sp^3 C–N coupling via dual copper and photoredox catalysis. Nature 559:83–88. (d) Hu X, Zhang G, Bu F, Nie L, Lei A (2018) Electrochemical-oxidation-induced site-selective intramolecular C(sp^3)–H amination. ACS Catal 8:9370–9375. (e) Xiang J, Shang M, Kawamata Y, Lundberg H, Reisberg SH, Chen M, Mykhailiuk P, Beutner G, Collins MR, Davies A, Bel MD, Gallego GM, Spangler JE, Starr J, Yang S, Blackmond DG, Baran PS (2019) Hindered dialkyl ether synthesis with electrogenerated carbocations. Nature 573:398–402. (f) Shibutani S, Kodo T, Takeda M, Nagao K, Tokunaga N, Sasaki Y, Ohmiya H (2020) Organophotoredox-catalyzed decarboxylative C(sp^3)–O bond formation. J Am Chem Soc 142:1211–1216. (g) Wang H, Liang K, Xiong W, Samanta S, Li W, Lei A (2020) Electrochemical oxidation-induced etherification via C(sp^3)–H/O–H cross-coupling. Sci Adv 6:eaaz0590. (h) Lee BJ, DeGlopper KS, Yoon TP (2020) Site-selective alkoxylation of benzylic C–H bonds by photoredox catalysis. Angew Chem Int Ed 59:197–202. (i) Oliva M, Coppola GA, Van der Eycken EV, Sharma UK (2021) Photochemical and electrochemical strategies towards benzylic C–H functionalization: a recent update. Adv Synth Catal 363:1810–1834
5. For review, see: (a) Leonori D, Aggarwal VK (2014) Lithiation–borylation methodology and its application in synthesis. Acc Chem Res 47:3174–3183. (b) Fyfe JWB, Watson AJB (2017) Recent developments in organoboron chemistry: old dogs, new tricks. Chemistry 3:31–55. (c) Friese FW, Studer A (2019) New avenues for C-B bond formation via radical intermediates. Chem Sci 10:8503–8518. (d) Wang M, Shi Z (2020) Methodologies and strategies for selective borylation of C–Het and C–C bonds. Chem Rev 120:7348–7398. (e) Das KK, Paul S, Panda S (2020) Transition metal-free synthesis of alkyl pinacol boronates. Org Biomol

Chem 18:8939–8974. (f) Das KK, Panda S (2020) Functionalization of heterocycles through 1,2-metallate rearrangement of boronate complexes. Chem Eur J 26:14270–14282. (g) Manna S, Das KK, Nandy S, Aich D, Paul S, Panda S (2021) A new avenue for the preparation of organoboron compounds via nickel catalysis. Coord Chem Rev 448:214165. (h) Lai D, Ghosh S, Hajra A (2021) Light-induced borylation: developments and mechanistic insights. Org Biomol Chem 19:4397–4428. (i) Kanti Das K, Manna S, Panda S (2021) Transition metal catalyzed asymmetric multicomponent reactions of unsaturated compounds using organoboron reagents. Chem Commun 57:441–459. Representatively, the following protocols were utilized to prepare the alkylboron substrates used in this study (j) Clary JW, Rettenmaier TJ, Snelling R, Bryks W, Banwell J, Wipke WT, Singaram B (2011) Hydride as a leaving group in the reaction of pinacolborane with halides under ambient grignard and barbier conditions. One-pot synthesis of alkyl, aryl, heteroaryl, vinyl, and allyl pinacolboronic esters. J Org Chem 76:9602–9610. (k) Roesner S, Brown CA, Mohiti M, Pulis AP, Rasappan R, Blair DJ, Essafi S, Leonori D, Aggarwal VK (2014) Stereospecific conversion of alcohols into pinacol boronic esters using lithiation–borylation methodology with pinacolborane. Chem Commun 50:4053–4055. (l) Atack TC, Cook SP (2016) Manganese-catalyzed borylation of unactivated alkyl chlorides. J Am Chem Soc 138:6139–6142. (m) Li C, Wang J, Barton LM, Yu S, Tian M, Peters DS, Kumar M, Yu AW, Johnson KA, Chatterjee AK, Yan M, Baran PS (2017) Decarboxylative borylation. Science 356:eaam7355. (n) Yang Y, Tsien J, Ben David A, Hughes JME, Merchant RR, Qin T (2021) Practical and modular construction of C(sp^3)–rich alkylboron compounds. J Am Chem Soc 143:471–480. (o) Yang Y, Tsien J, Hughes JME, Peters BK, Merchant RR, Qin T (2021) An intramolecular coupling approach to alkyl bioisosteres for the synthesis of multisubstituted bicycloalkylboronates. Nat Chem 13:950–955

6. (a) Brown HC, Rao BCS (1956) A new technique for the conversion of olefins into organoboranes and related alcohols. J Am Chem Soc 78:5694–5695. (b) Matteson DS, Mah RWH (1963) Neighboring boron in nucleophilic displacement. J Am Chem Soc 85:2599–2603. (c) Zweifel G, Arzoumanian H, Whitney CC (1967) A convenient stereoselective synthesis of substituted alkenes via hydroboration–iodination of alkynes. J Am Chem Soc 89:3652–3653. (d) Aggarwal VK, Fang GY, Schmidt AT (2005) Synthesis and applications of chiral organoboranes generated from sulfonium ylides. J Am Chem Soc 127:1642–1643
7. (a) Brown HC, Jadhav PK (1983) Asymmetric carbon–carbon bond formation via *B*-allyldiisopinocampheylborane. Simple synthesis of secondary homoallylic alcohols with excellent enantiomeric purities. J Am Chem Soc 105:2092–2093. (b) Miyaura N, Ishiyama T, Ishikawa M, Suzuki A (1986) Palladium-catalyzed cross-coupling reactions of B-alkyl-9-BBN or trialkylboranes with aryl and 1-alkenyl halides. Tetrahedron Lett 27:6369–6372. (c) Molander GA, Ito T (2001) Cross-coupling reactions of potassium alkyltrifluoroborates with aryl and 1-alkenyl trifluoromethanesulfonates. Org Lett 3:393–396
8. (a) Molander GA, Colombel V, Braz VA (2001) Direct alkylation of heteroaryls using potassium alkyl- and alkoxymethyltrifluoroborates. Org Lett 13:1852–1855. (b) Yasu Y, Koike T, Akita M (2012) Visible Light-induced selective generation of radicals from organoborates by photoredox catalysis. Adv Synth Catal 354:3414–3420. (c) Tellis JC, Primer DN, Molander GA (2014) Single-electron transmetalation in organoboron cross-coupling by photoredox/nickel dual catalysis. Science 345:433. (d) Duret G, Quinlan R, Bisseret P, Blanchard N (2015) Boron chemistry in a new light. Chem Sci 6:5366–5382. (e) Lennox AJJ, Nutting JE, Stahl SS (2018) Selective electrochemical generation of benzylic radicals enabled by ferrocene-based electron-transfer mediators. Chem Sci 9:356–361
9. Reported examples using oxidative conditions in chemical manners: (a) Cazorla C, Métay E, Lemaire M (2011) Oxidative nucleophilic substitution: transformation of alkylboronic derivatives. Tetrahedron 67:8615–8621. (b) Grayson JD, Dennis FM, Robertson CC, Partridge BM (2021) Chan–Lam amination of secondary and tertiary benzylic boronic esters. J Org Chem 86:9883–9897. Reported examples using oxidative conditions in electrochemical manners: (c) Suzuki J, Tanigawa M, Inagi S, Fuchigami T (2016) Electrochemical oxidation of organotrifluoroborate compounds. ChemElectroChem 3:2078–2083. (d) Tanigawa M, Kuriyama Y, Inagi S, Fuchigami T (2016) Electrochemical properties and reactions of sulfur-containing

organoboron compounds. Electrochim Acta 199:314–318. (e) Inagi S, Fuchigami T (2017) Electrochemical properties and reactions of organoboron compounds. Curr Opin Electrochem 2:32–37. (f) Ohtsuka K, Inagi S, Fuchigami T (2017) Electrochemical properties and reactions of oxygen-containing organotrifluoroborates and their boronic acid esters. ChemElectroChem 4:183–187

10. For review, see (a) Sperry JB, Wright DL (2006) The application of cathodic reductions and anodic oxidations in the synthesis of complex molecules. Chem Soc Rev 35:605–621. (b) Francke R, Little RD (2014) Redox catalysis in organic electrosynthesis: basic principles and recent developments. Chem Soc Rev 43:2492–2521. (c) Yan M, Kawamata Y, Baran PS (2017) Synthetic organic electrochemical methods since 2000: on the verge of a renaissance. Chem Rev 117:13230–13319. (d) Yan M, Kawamata Y, Baran PS (2018) Synthetic organic electrochemistry: calling all engineers. Angew Chem Int Ed 57:4149–4155. (e) Tang S, Liu Y, Lei A (2018) Electrochemical oxidative cross-coupling with hydrogen evolution: a green and sustainable way for bond formation. Chem 4:27–45. (f) Wiebe A, Gieshoff T, Möhle S, Rodrigo E, Zirbes M, Waldvogel SR (2018) Electrifying organic synthesis. Angew Chem Int Ed 57:5594–5619. (g) Kingston C, Palkowitz MD, Takahira Y, Vantourout JC, Peters BK, Kawamata Y, Baran PS (2020) A survival guide for the "electro-curious". Acc Chem Res 53:72–83. (h) Little RD (2020) A perspective on organic electrochemistry. J Org Chem 85:13375–13390. (i) Zhu C, Ang NWJ, Meyer TH, Qiu Y, Ackermann L (2021) Organic electrochemistry: molecular syntheses with potential. ACS Cent Sci 7:415–431. (j) Yuan Y, Yang J, Lei A (2021) Recent advances in electrochemical oxidative cross-coupling with hydrogen evolution involving radicals. Chem Soc Rev 50:10058–10086. For selected recent examples, see: (k) Sauermann N, Mei R, Ackermann L (2018) Electrochemical C–H amination by cobalt catalysis in a renewable solvent. Angew Chem Int Ed 57:5090–5094. (l) Gao X, Wang P, Zeng L, Tang S, Lei A (2018) Cobalt(II)-catalyzed electrooxidative C–H amination of arenes with alkylamines. J Am Chem Soc 140:4195–4199
11. Molander GA, Sandrock DL (2009) Potassium trifluoroborate salts as convenient, stable reagents for difficult alkyl transfers. Curr Opin Drug Discov Devel 12:811–823
12. During the preparation of the manuscript, a related electrochemical activation of $C(sp^2)$–B bond was reported. Wigman B, Lee W, Wei W, Houk KN, Nelson HM (2022) Electrochemical fluorination of vinyl boronates through donor-stabilized vinyl carbocation intermediates. Angew Chem Int Ed 61:e202113972
13. For a review on electrochemical examination of organic reactions, see: (a) Sandford C, Edwards MA, Klunder KJ, Hickey DP, Li M, Barman K, Sigman MS, White HS, Minteer SD (2019) A synthetic chemist's guide to electroanalytical tools for studying reaction mechanisms. Chem Sci 10:6404–6422. For recent examples, see: (b) Peters BK, Rodriguez KX, Reisberg SH, Beil SB, Hickey DP, Kawamata Y, Collins M, Starr J, Chen L, Udyavara S, Klunder K; Gorey TJ, Anderson SL, Neurock M, Minteer SD, Baran PS (2019) Scalable and safe synthetic organic electroreduction inspired by li-ion battery chemistry. Science 363:838. (c) Hu P, Peters BK, Malapit CA, Vantourout JC, Wang P, Li J, Mele L, Echeverria P-G, Minteer SD, Baran PS (2020) Electroreductive olefin–ketone coupling. J Am Chem Soc 142:20979–20986
14. Shin SJ, Kim JY, An S, Kim M, Seo M, Go SY, Chung H, Lee M, Kim M-G, Lee HG, Chung TD (2022) Revisiting thin-layer electrochemistry in a chip-type cell for the study of electro-organic reactions. Anal Chem 94:1248–1255
15. Kotsinaris A, Kyriacou G, Lambrou CH (1998) Electrochemical reduction of dichloromethane to higher hydrocarbons. J Appl Electrochem 28:613–616
16. Chen N, Ye Z, Zhang F (2021) Recent progress on electrochemical synthesis involving carboxylic acids. Org Biomol Chem 19:5501–5520
17. Li Y, Vaz RJ, Olson SH, Munson M, Paras NA, Conrad J (2020) Selectivity in the addition of electron-deficient radicals to the *C2* position of indoles. Eur J Org Chem 2020:5828–5832
18. Taylor AP, Robinson RP, Fobian YM, Blakemore DC, Jones LH, Fadeyi O (2016) Modern advances in heterocyclic chemistry in drug discovery. Org Biomol Chem 14:6611–6637
19. (a) Hammett LP (1937) The effect of structure upon the reactions of organic compounds. Benzene derivatives. J Am Chem Soc 59:96–103. (b) Hansch C, Leo A, Taft RW (1991)

A survey of Hammett substituent constants and resonance and field parameters. Chem Rev 91:165–195

20. (a) Della EW, Grob CA, Taylor DK (1994) Bridgehead carbocations: A solvolytic study of 1-bromobicyclo[1.1.1]pentane and its bridgehead-substituted derivatives. J Am Chem Soc 116:6159–6166. (b) Livesley S, Sterling AJ, Robertson CM, Goundry WRF, Morris JA, Duarte F, Aïssa C (2022) Electrophilic activation of [1.1.1]propellane for the synthesis of nitrogen-substituted bicyclo[1.1.1]pentanes. Angew Chem Int Ed 61:e202111291
21. Sheng T, Zhang H-J, Shang M, He C, Vantourout JC, Baran PS (2020) Electrochemical decarboxylative *N*-alkylation of heterocycles. Org Lett 22:7594–7598
22. Bagutski V, Ros A, Aggarwal VK (2009) Improved method for the conversion of pinacolboronic esters into trifluoroborate salts: facile synthesis of chiral secondary and tertiary trifluoroborates. Tetrahedron 65:9956–9960
23. Roesner S, Brown CA, Mohiti M, Pulis AP, Rasappan R, Blair DJ, Essafi S, Leonori D, Aggarwal VK (2014) Stereospecific conversion of alcohols into pinacol boronic esters using lithiation-borylation methodology with pinacolborane. Chem Commun 50:4053–4055
24. Li C, Wang J, Barton LM, Yu S, Tian M, Peters DS, Kumar M, Yu AW, Johnson KA, Chatterjee AK, Yan M, Baran PS (2017) Decarboxylative borylation. Science 356:eaam7355
25. Clary JW, Rettenmaier TJ, Snelling R, Bryks W, Banwell J, Wipke WT, Singaram B (2011) Hydride as a leaving group in the reaction of pinacolborane with halides under ambient Grignard and Barbier conditions. One-pot synthesis of alkyl, aryl, heteroaryl, vinyl, and allyl pinacolboronic esters. J Org Chem 76:9602–9610
26. Yang Y, Tsien J, Hughes JME, Peters BK, Merchant RR, Qin T (2021) An Intramolecular coupling approach to alkyl bioisosteres for the synthesis of multisubstituted bicycloalkylboronates. Nat Chem 13:950–955
27. Atack TC, Cook SP (2016) Manganese-catalyzed borylation of unactivated alkyl chlorides. J Am Chem Soc 138:6139–6142
28. Yang Y, Tsien J, Ben David A, Hughes JME, Merchant RR, Qin T (2021) Practical and modular construction of C(sp^3)-rich alkylboron compounds. J Am Chem Soc 143:471–480
29. Wang W, Ding C, Yin G (2020) Catalyst-controlled enantioselective 1,1-arylboration of unactivated olefins. Nat Catal 3:951–958
30. Sorin G, Martinez Mallorquin R, Contie Y, Baralle A, Malacria M, Goddard J-P, Fensterbank L (2010) Oxidation of alkyl trifluoroborates: an opportunity for tin-free radical chemistry. Angew Chem Int Ed 49:8721–8723
31. Huang H, Zhang G, Gong L, Zhang S, Chen Y (2014) Visible-light-induced chemoselective deboronative alkynylation under biomolecule-compatible conditions. J Am Chem Soc 136:2280–2283
32. Dai J-J, Zhang W-M, Shu Y-J, Sun Y-Y, Xu J, Feng Y-S, Xu H-J. Deboronative cyanation of potassium alkyltrifluoroborates via photoredox catalysis. Chem Commun 52:6793–6796
33. Hammett LP (1937) The effect of structure upon the reactions of organic compounds. Benzene derivatives. J Am Chem Soc 59:96–103
34. Hansch C, Leo A, Taft RW (1991) A survey of Hammett substituent constants and resonance and field parameters. Chem Rev 91:165–195
35. Tsukada H, Mukaeda Y, Hosokawa S (2013) *syn*-selective Kobayashi aldol reaction using acetals. Org Lett 15:678–681
36. Lee YH, Morandi B (2017) Ether synthesis through reductive cross-coupling of ketones with alcohols using Me_2SiHCl as both reductant and Lewis acid. Chem Commun 28:2425–2428
37. Barluenga J, Tomás-Gamasa M, Aznar F, Valdés C (2010) Straightforward synthesis of ethers: metal-free reductive coupling of tosylhydrazones with alcohols or phenols. Angew Chem Int Ed 49:4993–4996
38. Xiang J, Shang M, Kawamata Y, Lundberg H, Reisberg SH, Chen M, Mykhailiuk P, Beutner G, Collins MR, Davies A, Del Bel M, Gallego GM, Spangler JE, Starr J, Yang S, Blackmond DG, Baran PS (2019) Hindered dialkyl ether synthesis with electrogenerated carbocations. Nature 573:398–402

39. Yamada K, Hayakawa N, Fujita H, Kitamura M, Kunishima M (2016) Development of a triazine-based *tert*-butylating reagent, TriAT-*t*Bu. Eur J Org Chem 2016:4093–4098
40. Vasilopoulos A, Golden DL, Buss JA, Stahl SS (2020) Copper-catalyzed C–H fluorination/functionalization sequence enabling benzylic C-H cross coupling with diverse nucleophiles. Org Lett 22:5753–5757
41. Gawande MB, Rathi AK, Branco PS, Nogueira ID, Velhinho A, Shrikhande JJ, Indulkar UU, Jayaram RV, Ghumman CAA, Bundaleski N, Teodoro OMND (2012) Regio- and chemoselective reduction of nitroarenes and carbonyl compounds over recyclable magnetic ferrite-nickel nanoparticles (Fe_3O_4-Ni) by using glycerol as a hydrogen source. Chem Eur J 18:12628–12632
42. Longwitz L, Jopp S, Werner T (2019) Organocatalytic chlorination of alcohols by P(III)/P(V) redox cycling. J Org Chem 84:7863–7870
43. Jiang W, Li N, Zhou L, Zeng Q (2018) Copper-catalyzed stereospecific C–S coupling reaction of enantioenriched tertiary benzylic amines via in situ Activation with methyl triflate. ACS Catal 8(11):9899–9906
44. Liu S, Jin S, Wang H, Qi Z, Hu X, Qian B, Huang H (2021) Nickel-catalyzed oxidative dehydrogenative coupling of alkane with thiol for C(sp^3)-S bond formation. Tetrahedron Lett 68:152950
45. SMART, SAINT and SADABS (2016) Bruker AXS Inc., Madison, Wisconsin, USA
46. Sheldrick GM (2002) SADABS v 2.03, University of Göttingen, Germany
47. SHELXTL v 6.14 (2000) Bruker AXS, Inc., Madison, Wisconsin, USA
48. Toto P, Retailleau P, Boivin J (2016) CSD communication. https://doi.org/10.5517/ccdc.csd.cc1n5f3c